Natural Stain Remover

Note on measurements, ingredients and equipment.
Exact weights and measures are not necessary for the 'recipes' in this book. General proportions are sufficient. A cup or glass, some teaspoons and tablespoons and a bucket are generally all you will need.
A measuring cup equals 250 ml (8 fl oz) but any cup size will do; for example, a standard British teacup. If your cup is smaller than 250 ml, adjust the spoon measures accordingly. Keep a designated cup or glass aside for cleaning purposes and do not use it for cooking.
The ingredients in this book are common household items and are readily available in the United Kingdom.

Natural Stain Remover

Natural Stain Remover

clean your home without harmful chemicals

ANGELA MARTIN

APPLE

Contents

Introduction

Here's one of the ironies of running a home: while you are working at keeping your immediate world clean for yourself and your family, you could be compromising everyone's health and the health of the environment.

You may keep your floors scrubbed, your living areas dust-free, your kitchen surfaces germ-free, your furniture highly polished and your clothing wellwashed and stain-free, but what are you using to achieve this? If it's nonorganic commercial cleaning products you're relying on, then in trying to solve one potential problem you may be introducing another.

Check out the average house, and you'll find any number of hazardous chemicals lurking:

- in kitchen cupboards
- in bathroom cabinets
- on laundry shelves
- in garden sheds

They may come in bright, attractive packages, and their labels may present images of cleanliness and purity, but the products themselves may also contain substances that would disturb you — if only you knew their nature and effects. For example, did you know that a solvent found in many common commercially available nonorganic detergents has been linked to respiratory illness?

Every day, the environment is being penetrated by loads of chemicals present in cleaning products used around the home. These chemicals are not the only cause of polluted air or waterways, but they are a contributing factor. What's more, we have the choice of whether or not we add to the use of these chemicals.

If you are reading this book, it's probable that you have concerns about what is sprayed into the air, used near skin, or washed down the sink. You may want more direct input into what is being used around your family and your home. You may be wondering how you can help to achieve an ecologically sustainable future for yourself and your children.

Remember that every person makes a difference, and every small step does too. Once you recognize this, you'll be able to look for ways you can contribute. If the environment benefits, so will your family.

This book offers you many ways to make changes around the home. You'll be surprised at their simplicity, and amazed that all you need to keep your home sparkling are everyday products that are perfectly safe for the health of your family and your environment.

Why Organic?

THE DANGERS OF TOXIC PRODUCTS

There was a time when people used simple substances to clean up stains, disinfect, wash, and do the polishing around the home. These products seemed perfectly efficient, and their use was based on household wisdom, passed down from parent to child for centuries.

Before you label those times of simple products "the good old days," however, keep in mind that in those days, too, toxic ingredients were often present in the home. For example, although asbestos was once considered a miracle because of its highly desirable ability to resist fire, we are now aware of its devastating legacy. Likewise, the parents of children who playfully picked and chewed at peeling paint had no idea of the disastrous consequences for their children of ingesting the lead in that paint.

We are far better informed now. We know far more about the factors affecting our families' health. We also know far more about the ingredients in the products we bring home for daily use. With this increased knowledge comes responsibility, and the power to make sound choices.

THE RISE OF CHEMICALS

Before the 1940s, basic everyday products were used to keep the house clean. Even cosmetics were based on kitchen products such as eggs, oil and vinegar.

In the late 1930s, many newly developed chemicals were considered safe only because the skin was regarded as an effective barrier to such products. After World War II, research on chemicals proliferated, and householders were offered magical solutions to every kind of job throughout the house and in the garden. New products appeared to offer limitless possibilities for a "better" life.

Both increasing competition between brands, and increasingly large advertising campaigns, have boosted the use of these products — to the point where many householders now find it difficult to imagine a clean home environment without them. To watch this phenomenon in action, take a look at some of the commercials featuring cleaning products. Take note of the claims, and also the emphasis on whiteness and purity. Often the underlying message is that if you were truly clean and responsible, and cared for those around you, you would choose such a product for your home. We are led to believe that we will only be safe and clean, and our immediate environment sparkling and germ-free, when we use these commercially available products.

INDOOR POLLUTION

As the U.S. Environmental Protection Agency has pointed out, while most people are aware of the damage that outdoor air pollution can do to their health, they are less aware of the harmful effects of indoor pollution. In fact, studies have indicated that indoor levels of pollutants can be 25 times higher than outdoor levels, and, in some cases, 100 times higher.

The high exposure to indoor pollutants is due to a number of factors. For example, to save energy, buildings may sometimes be more tightly sealed, thereby reducing ventilation rates. The decision to take measures such as these may be out of our hands, but we can do other things instead to counter the high exposure rates.

We can make the decision to reduce or eliminate altogether the use of products based on chemical formulas: personal products, pesticides and household cleaners. If we avoid purchasing these, the air in our homes will be less polluted. And that will be good news for both our personal health and the health of the environment.

"Awareness of indoor air pollution as an environmental issue is relatively new. Indoor air pollutants can have both short-term and long-term effects on health, especially when concentrations build up. One challenge for researchers today is to increase our understanding of the possible health impacts of being exposed to mixtures of indoor air pollutants at low levels for long periods of time." U.S. Environmental Protection Agency, 1993

PERSONAL HEALTH

Cleaning products that contain certain chemicals may pose a health threat to us, and to those around us, because of the hazardous nature of their ingredients.

At least some of the chemicals contained in many popular household cleaners are poisonous, corrosive and flammable. Exposure to such chemicals can cause a huge range of reactions, such as headaches, lung irritation, irritability, fatigue, and even cause damage to the nervous system, the mucous membranes, the kidneys, liver, skin and eyes. The fumes from cleaning products can also trigger asthma attacks and other respiratory problems.

For some people, the effects of exposure are immediate. For others, it may be years before the effects are felt. Long or repeated exposure can have a cumulative effect. One challenge for researchers today is to increase our understanding of the possible health impacts of being exposed to mixtures of indoor air pollutants, at low levels, for long periods of time.

Children are far more vulnerable to pollutants of all types than adults. Their systems are still developing, and it takes far less of any product to affect a small child than a fully grown person. Children will also poke and scratch at areas where residues hide — and then put their fingers in their mouths.

Also, do not overlook the health of your pets. Their health may also be compromised if they are exposed to the toxins present in many chemicals.

THE ENVIRONMENT

Sometimes we talk as if the environment were something separate from ourselves, as if we can remain unaffected even as the environment is being degraded. But we too are being affected.

- The water we drink and the air we breathe are extremely vulnerable. It takes only small quantities of solvents and other chemicals to pollute the large amount of water we use. The chemicals found in modern cleaners find their way into rivers, lakes and groundwater. For example, phosphates, found in many cleaning products, accelerate the growth of algae and plants in natural waters, leading to the depletion of oxygen in the water mass and the loss of species. What's more, when algae forms in large amounts, the water beneath it can become putrid.
- The compounds released from some aerosol products are responsible for depleting ozone in the atmosphere.
- Toxic fumes from certain products add to the poison floating in the air, which we know as pollution.

Being alert to the ways that the environment is being degraded is the first step in helping turn the situation around. You can make a start now by cutting back on your own use of toxic products, and using instead the organic products suggested in this book.

THE BEAUTY OF CLEANING ORGANICALLY

WHAT IS "ORGANIC"?

The term "organic" usually refers to anything that is natural, unrefined and untreated. The ingredients offered in this book fall into this category, and on pages 17 to 25 you'll find a description of each of them. Their uses will be detailed in "Recipes: Formulas, blends and treatments" on pages 33 to 61, and in "The A–Z of stain removal and cleaning" on pages 62 to 117.

While organic products are natural, some of them do hold dangers. For example, borax is a marvelous stain remover and has many other uses as well, but although it is a naturally occurring mineral salt, it is poisonous and must not be used around children or pets.

However, this warning applies to very few of the organic products suggested for use in this book. Generally speaking, organic cleaning products are safer overall than nonorganic cleaning products. After all, some of them are even used in cooking!

THREE GOOD REASONS TO SWITCH

There are three main reasons for switching to organic cleaning products:

1. Your health and the health of those around you will benefit as exposure to toxic substances is reduced.

2. The environment will be less burdened and depleted.
3. You will spend less money, because the cost of organic cleaning products is usually far lower than the cost of the commercially available nonorganic variety.

There are other benefits too. For example, once you start using organic ingredients you'll find yourself wanting to experiment. You'll be doing your own mixing, and you may be doing your own creating too. You'll find yourself becoming more curious about how to strengthen their power, how to add natural perfumes, and how to adapt the products for specific situations. You will feel very resourceful, and this feeling will spill over into other aspects of your life as well.

The other benefit is that when your children see you using these basic household products to clean and remove stains, they will pick up on attitudes and approaches that show respect for the environment, both in the home and outside of it.

DO THEY WORK?

If you've spent years using nonorganic products to clean, shine and remove stains, then the organic products may take some getting used to. Nonorganic products often seem to achieve instant results with little effort. But remember, the ingredients in many of these products are dangerously strong, and that while you are using them, you are inhaling their toxic fumes and coming into contact with their damaging effects on the skin.

Are the fast results worth it? Note also, that when cleaning takes a bit longer with these products, it means longer exposure to chemicals that may be affecting your well-being.

You'll find that organic products will work just as well. You may have to apply a little elbow grease at times, but in many cases you'll be astounded at how quickly and almost effortlessly they remove stains.

Also keep in mind that basic products were being used for cleaning for a very long time before the Chemical Age began. It's worth reflecting on how much advertising has influenced us, encouraging us to buy a whole range of commercial nonorganic cleaning products. Practice replacing the images created by the commercials with your own images of healthy cleaning substances.

ONE STEP AT A TIME

Does changing from nonorganic cleaners to organic cleaners seem too big a jump? If you don't want to get rid of all your bottles and cartons and jars of nonorganic products in one go, then change one step at a time.

You could start with a simple paste of baking soda and water to remove stains from your kitchen counter and to clean your stainless steel sink — or a paste of cream of tartar and water to remove stains from fabric. You'll be amazed at how effective it is.

Many recipes for organic cleaners include small quantities of essential oils. Not only do these oils add perfume to your mixes, but they are also effective in their own right. For example, lemon oil and the other citrus oils will cut grease, and lavender oil has antifungal and antibacterial properties.

Try out the recipes in this book. They may well change your approach to cleaning. Once you test their effectiveness for yourself, you may want to expand the experience. Before long you may find yourself buying baking soda and other such organic cleaning products in bulk, rather than spending time in front of the enormous range of brightly packaged cleaners, trying to decide which to choose.

GOOD-BYE TO HAZARDOUS HOUSEHOLD PRODUCTS

THE PROBLEM OF STORAGE

The use of nonorganic chemical cleaners is concerning enough, but compounding the problem is the extent to which the average home accumulates and stores them. Around your home you could find a great deal of dangerous waste materials, some of which are cleaning agents, and the rest an accumulation of paints, chemicals and pesticides.

DISPOSING OF TOXIC PRODUCTS SAFELY

Be sure to dispose thoughtfully of products you no longer want to use. Those with hazard symbols should not be poured down the drain or put into the garbage. Leave them at a hazardous-waste facility. Contact your local municipal authorities for details of the nearest facility.

The basics of stain removal and cleaning

ORGANIC CLEANING PRODUCTS: A SHOPPING LIST

You will already have some of the items on this list on your shelves. For example, most kitchens contain baking soda, salt and vinegar, and often lemons too, as they are useful for many dishes and dressings. These four items alone are enough to use as cleaners. You could start with them, getting to know them and recognize their cleaning power, and then add the others from the list as you go. All items are readily available from supermarkets.

THE BASIC DOZEN

1. Baking soda
2. Borax
3. Cream of tartar
4. Eucalyptus oil
5. Glycerin
6. Lemons
7. Pure soap
8. Salt
9. Soda water
10. White vinegar
11. Washing soda
12. Essential oils

PRODUCT DESCRIPTION

Baking soda (bicarbonate of soda)

This is a powder that is mildly alkaline, and is therefore useful for neutralizing mild acids. It is excellent as a cleaner and stain remover, and also as a deodorizer. Its uses are many around the home: keep it handy in the kitchen, laundry and bathroom. You can:

- Clean burned food at the bottom of pots and pans with a paste of baking soda and water (for recipe see p102).
- Brighten old silver by dipping it into a solution of baking soda and boiling water (for recipe see p107).
- Eliminate old ink stains from fabric and carpets by applying a paste of baking soda and water.
- Add baking soda to white vinegar for a tough bathroom cleaner (for recipe see p41).

Borax

This is a mineral salt that occurs naturally. It is a combination of boric acid and soda, and has bleaching qualities. When dissolved in water, borax forms an alkaline antiseptic solution, and is useful as a disinfectant, detergent and water softener.

One of the great properties of borax is that it will loosen grease and dirt, making it a very effective stain remover. You can:

- Use borax in hot water to sponge away syrup stains in carpet and fabric (for recipe see p110).
- Add borax to baking soda in a shaker to make a quick and excellent cleaner for the bathroom (for recipe see p42).

- Make up a spray of borax, glycerin and liquid castile soap to apply to stubborn stains (for recipe see p58).
- Remove stains from a sheepskin or lambskin rug by sponging with a solution of borax and warm water (for recipe see p106).

NOTE

- *Borax is not the same as boric acid.*
- *Borax is poisonous. Don't use it around children or pets.*
- *Wear rubber gloves when using borax, because it can irritate the skin and eyes, and will enter the body through broken or cracked skin.*

Cream of tartar

This is a mildly acidic powder, commonly combined with baking soda to make what is known as baking powder, a leavening (raising) agent. It was first made in the early 1800s, from the juice of fruits such as grapes, pineapples and mulberries.

Cream of tartar is another excellent stain remover. You can:

- Mix it to a paste with water to apply to the ring around shirt collars (for recipe see p77).
- Sprinkle it onto rust spots on fabric and then dip into hot water to remove the rust (for recipe see p104).

Eucalyptus oil

This is an oil with a distinctive smell distilled from the leaves of Australia's eucalyptus trees. It is useful as a disinfectant, and in fact is often an ingredient in household disinfectant products, as well as in detergent solutions for washing woolen fabrics. As a bonus, eucalyptus oil is also a natural insect repellent.

It is an effective agent in stain removal, doing those jobs that few other products can do. You can:

- Dab eucalyptus oil onto adhesive labels to remove them (for recipe see p64).
- Apply a few drops to a damp cloth to give a bath, especially a dark bath, a great shine.
- Use it to treat bloodstains on both fabric and carpet (for recipe see p70).
- Clean rust stains on chrome with the oil before rubbing them away with aluminum foil.

NOTE

Keep eucalyptus oil away from children. It is highly toxic when swallowed.

Glycerin

Glycerin is a colorless, odorless, syrupy liquid. You may have it in your kitchen cupboard for use in confectionery, and it is also used in frosting (icing) to prevent it from hardening excessively. But it has other uses too. Glycerin is good for loosening some types of stains. You can:

- Rub glycerin into very tough stains to soften and release the stain before presoaking or laundering (for recipe see p58).
- Dab glycerin into newsprint stains to loosen the print (for recipe see p98).
- Use glycerin to remove perfume stains that persist in fabric or carpet (for recipe see p101).

Lemons

Lemon juice is known as nature's own bleach and disinfectant. It is the acidic character of the lemon that makes its juice so useful as a mild bleach and cleaning agent. Its efficiency and strength are enhanced when used with baking soda. In many instances, you can substitute the bottled variety if you have no fresh lemons to squeeze. You can:

- Add lemon juice to baking soda or borax for extra cleaning power.
- Sprinkle half a cut lemon with salt, and use this to clean away stains on many metals.
- Use lemon juice with salt and sun to remove rust stains from fabric (for recipe see p104).
- Clean away tea stains by applying a little lemon juice before washing as usual (for recipe see p111).

Pure soap

You can base your washing powders and liquids, whether in the kitchen or in the laundry, on pure soap. Look out for soap flakes, or grate your own flakes from bars of pure soap.

Castile soap is a pure soap that can be bought in health-food and natural-product stores in solid or liquid form. Its name refers to the area in Spain where the soap originated. A mild soap made with olive oil and coconut oil, it is especially good for dry and sensitive skins, is biodegradable, and contains no animal fats or synthetic detergents. You can:

- Add it to glycerin and borax for a pretreatment spray in the laundry (for recipe see p58).
- To wash baby clothes, add it to warm water, with a few drops of essential oil of your choice (for recipe see p61).
- Use it to wash delicate clothing such as lingerie (for recipe see p61).

Salt

This white crystalline powder has an abrasive action, and is particularly effective when used with lemon juice. You can:

- Soak egg-stained fabric in cold water and salt to remove the stain (for recipe see p82).
- Use salt and lemon juice to clean away soap buildup from glass shower doors (for recipe see p86).
- Clean up food spills in the oven by sprinkling liberally with salt while the oven is still warm. Clear away the debris a few hours later.

Soda water

This must be one of the best emergency products around. Keep bottles handy throughout your home. Pour or spray soda water onto a fresh stain as soon as possible. The carbonation — the fizziness — will bubble the staining substance to the surface, allowing you to blot it away. There is also salt in soda water, which will help prevent permanent staining.

Washing soda

This is a crystalline powder that makes up the greater part of most commercially available nonorganic laundry formulas. You can make your own organic laundry cleaner with washing soda and other natural ingredients for a fraction of the cost. You can:

- Use washing soda as a base to make up laundry powders or liquids that will clean your clothes naturally (for recipe see pp59–60).
- Add borax and white vinegar to washing crystals to make a laundry powder for hard-water areas.

White vinegar

Ordinary white vinegar from the supermarket is a central ingredient in many kinds of cuisine, and is equally versatile as a cleaning agent. It is mildly acidic, so it has a similar effect to lemon juice. It can neutralize grease, fight mold, act as a disinfectant, bleach, and deodorize. You can:

- Remove alcohol stains from fabric by soaking in vinegar and cold water for a couple of hours (for recipe see p65).
- To clean and sanitize, wipe over the bathroom with a cloth dipped in white vinegar.
- Use heated white vinegar to clean away paint spots on windows and mirrors.

NOTE

When heating vinegar heat to very warm or hot—either in a saucepan on the stove, or in the microwave (1 cup on high for 1 minute).

Essential oils

Think of having your favorite fragrance wafting from the bathroom, the kitchen, and every other room — nothing artificial, just the exquisite scent of an essential oil.

It isn't just their scent that makes these many oils so good to use. They have important properties of their own, and are therefore excellent additions to the list of basic home cleaners. Many, such as lavender, citronella, pine, lemon and spearmint, are antibacterial. Others, such as eucalyptus, patchouli, sandalwood, lavender and lemon, are antifungal.

One of the beauties of making up your own cleaning products is that you can scent them in any way you choose. You'll find the commercial nonorganic varieties offer a very limited range of perfume. What's more, they can be overpowering, and may make you sneeze and splutter every time you use them.

Keep a few of your favorite oils at home, build on them as the weeks go by, and you can vary the scent through the house from room to room and from week to week.

Start with these fragrant possibilities:

- Lemon oil
- Lavender oil
- Rose oil
- Citronella oil
- Eucalyptus oil
- Orange oil
- Tea tree oil

Choose the ones that appeal to you and then broaden your range so you can vary your choice of fragrance to suit your mood and the season.

When you are buying essential oils, go for the pure, undiluted form. Pure oils come in a dark glass bottle, and are meant to be stored away from heat and direct light. You need only a few drops: remember that these are highly concentrated oils, and a little goes a long way.

In "Recipes: Formulas, blends and treatments," pages 33 to 61, you'll find that many of the recipes suggest combinations of essential oils. However, you can experiment with your own combinations. Here are some ideas to get you going:

This essential oil	blends well with
Lemon	chamomile, eucalyptus, lavender, rose, sandalwood, ylang-ylang, other citrus oils
Lavender	cedarwood, clary sage, patchouli, all citrus and floral oils
Rose	bergamot, chamomile, clary sage, geranium, jasmine, patchouli
Citronella	lavender, rose, geranium, basil, thyme, orange
Eucalyptus	cedarwood, lavender, lemon, pine, rosemary, thyme
Orange	all floral and citrus oils
Tea tree	clary sage, geranium, lavender, pine, rosemary, spice (e.g., nutmeg)

CAUTION

Never let children handle essential oils. They are strong, and can irritate the skin. They must be diluted with other liquid, as in the recipes in this book, before they are used, and it is recommended that you wear rubber gloves when handling formulas with essential oils as ingredients.

GETTING ORGANIZED

FIRST THINGS FIRST

You will find that you already have many of the tools you need for organic cleaning in your home. The list of basic equipment below covers the essentials of cleaning, but there may be other items you like that are not listed here. Use your imagination when collecting your materials, and bear in mind that any job will be easy if you have just the right tools.

You may like to keep all your cleaning gear in a single, large container in your laundry. This may be the room you choose for mixing all your formulas for use around the home. However, you may prefer to keep cleaning agents appropriate for particular jobs in the rooms of your home most likely to need them: the bathroom, laundry and kitchen. This way everything you need will be handy, especially for quick jobs like wiping the sink.

It's important to be organized from the beginning so that starting out on making and using your own organic cleaners will be an uncomplicated task. It will be all the more enjoyable for you if you've decided how to store materials and supplies, and where to store them.

THE BASIC EQUIPMENT

Brushes

- Hard-bristled brushes are essential for those scrubbing jobs where stubborn stains need a bit more elbow grease and the surface won't scratch easily. An example is bathroom tiles that need the occasional scrub to remove the buildup of soap scum. Keep a second hard-bristled brush for dry jobs, such as brushing away baking soda (used for stain removal) from sofa covers.
- A soft-bristled brush is good for jobs involving surfaces that scratch easily. Also, brushes with

soft and therefore flexible bristles will reach into corners that gather dust and dirt.

- Toothbrushes are great for those hard-to-get-at places, such as around the tap area, or the grooves of more ornate pieces. They're handy for small jobs where a larger brush would get in the way, and because of the long, thin handle, you can reach where other brushes and cloths cannot. The toothbrush that's getting a bit shaggy for your teeth is perfect for the cleaning kit — pop it in along with your other tools.
- Don't be tempted to go without a toilet brush: a quick brush with any number of simple, organic formulas will keep the toilet stain-free and disinfected.

Buckets

Buckets are useful for soaking items and for dipping into as you wash down or mop floors. Keep at least two, perhaps a large one and a smaller one, so that when one is in use, the other will be free. If you want a bucket you can carry around while cleaning, make it a smaller one — otherwise you'll find that, once you've filled it with liquid, it could be extremely heavy.

Cloths and towels

Cloths are the absolute essentials for cleaning. While they are easy to buy in packs and often quite inexpensive, you can make your own from clothing or sheeting that's seen better days. Cotton cloths are lovely to use, especially when they come from old T-shirts and underwear — these are soft and absorbent. Wildly colored cloths are lots of fun, but it may be wise to stick to white if you aren't sure about how colorfast the colored ones are. If not, you may end up staining the very area you are using the cloth to clean.

There's another good reason to have plenty of cloths handy around the home. Doing so will save on paper towels, often used for quick mop-ups and other jobs.

Towels that are fraying around the edges are also perfect for jobs around the home, especially when you need to absorb a spill quickly. Use them with soda water to mop up just about any stain. Keep most of your towels in one piece for those mopping-up jobs, and cut a few into smaller pieces for jobs that require those sizes.

Containers

You'll need various containers for storing basic ingredients, and the formulas and recipes given in this book. Keep an eye out for suitable sizes. As you pack away your groceries, take note of the containers you're emptying. Many of them will make perfect containers for cleaning formulas later on:

- Small jars are perfect for mixing up paste for spot cleaning.
- Shakers will be needed for jobs that require shaking out a powdery cleaner. Punch holes in the lid of any suitable container to make them.
- You can continue using the big plastic containers that used to hold your nonorganic laundry detergent — this time for your homemade variety.
- You'll also need spray bottles and squirt bottles. Again, you may have these already.

Make sure you clean them before use. You can also purchase both types of bottles.

Be sure to wash all containers thoroughly, rinse in white vinegar and dry well — preferably in the sun — before using.

Other useful items

For measuring out quantities, it's a good idea to have:

- A nest of cups, for measuring out 1/4 cup, 1/2 cup, and 1 cup; look for the kind usually used for baking.
- A funnel, if you are going to be making up formulas in bottles with narrow necks such as spray bottles.
- Spoons — a teaspoon and a tablespoon.
- A dropper for each essential oil, unless the bottle has the capacity to "pour" a drop at a time.

You'll also find these useful

- Rubber gloves, to protect your hands from some of the stronger products. Keep in mind that although the products used in this book are natural, they still may not be safe to handle without protective gloves. Borax, for example, has many uses around the home, yet it can make the skin peel and itch. Wear rubber gloves whenever you mix up formulas — even essential oils can be strong when they come into contact with your skin.
- A mop. This isn't essential, and some people prefer to use cloths, even for washing the floor. But it will be handy for larger areas like flooring, and for spots that you may not be able to reach without a ladder.

THE RULES OF STAIN REMOVAL

There are some important rules to follow when it comes to treating a stain, whatever the type. Keep a copy of this page in a place where you can refer to it quickly. If you stick to these rules, you will find organic stain removal both easy and effective.

RULE 1: Treat the stain as soon as possible. The sooner you apply treatment, the greater your chance of success, because "fresh" stains will have not had time to set.

RULE 2: Be patient. You may have to reapply the treatment several times before the stain comes away. With each treatment, a little more of the stain will be removed.

RULE 3: Avoid ironing over a stain, or putting stained clothes into the clothes drier. You will simply bake the stain in, making it virtually impossible to remove completely.

RULE 4: Blot, don't rub, in almost all cases. Rubbing causes fabric to become roughened, and may even damage it badly, causing it to rip. It also tends to spread the stain so that a larger area is affected. It is only occasionally that rubbing is recommended because the nature of the fabric and the source of the stain allow it.

RULE 5: Work from the outside of the stain toward the center. This way you'll avoid the telltale outer ring. This applies as much to clothing as it does to carpet and upholstery.

RULE 6: Don't be tempted to overlook spills because they appear to be colorless. Some, like the colorless stains left by soft drinks, will eventually turn brown because the sugar content will caramelize over time.

MORE ABOUT STAIN REMOVAL

A VARIETY OF METHODS

Stains can be treated and removed in a variety of ways, depending on the substance doing the staining, the substance stained, and the age of the stain. For some stains, however, you may only need to apply a single ingredient, and then leave it to do its work. For example, you can pour soda water onto certain carpet stains, allowing the fizz to lift the stain to the surface, and then press a towel onto the area to absorb the liquid.

In other cases, you may need to use a little fine steel wool to give an abrasive action to stubborn stains. You can also try using a toothbrush to reach hard-to-get-at places, or to treat embossed areas.

DIFFERENT MATERIALS

Whatever type of stain you are treating, always bear the material you are cleaning in mind. For example, it may not be worth trying to remove stains from very old fabric. Age may have worn it to the extent that the application of any cleaning agent, however mild, may damage the already weakened fibers.

Most clothing items come with washing instructions. Use these as a guide to the toughness of the fabric, and apply cleaning methods accordingly.

HELPFUL HINTS

- If you find a stain while you are ironing an article of clothing and you need to wear that article NOW, iron as close to the stain as possible without actually making contact with it. That way, you'll avoid setting it with the heat of the iron. Remember to treat the stain later on.
- The sun and the air are the best way to dry clothes. The sun will help whiten and the air will deodorize. If you are in the habit of using a clothes drier, be aware that it is an energy drain, and also that the drier will bake on any stains that have not yet been removed and make them even harder to remove later on.
- Test whether the fabric you are working on is colorfast. Take a white cotton cloth, dampen it, and then press it against an inconspicuous part of the fabric. If color comes away on the cloth, then you know that the color will run, and you will need to take special care when applying treatment for stains.

There are certain tasks around the home that, if undertaken regularly, will reduce your maintenance work in the long term. An example is the shower: if it is wiped down with white vinegar each day after use, you'll avoid the accumulation of soap scum that means a very tough job further down the line.

Recipes: formulas, blends and treatments

FIRST, LET'S TAKE A WALK

Yes, that's right, let's take a walk through the rooms and areas of your home. Take pen and paper with you, and jot down all the potential cleaning jobs you see. Look for grubby corners, stained surfaces, soiled spots, greasy marks, metals in need of a polish, crevices that need a clean, and mirrors and glass that are spotted and cloudy.

Jot down every job you'd like to see done. Then use this book to help you categorize what surfaces need cleaning, and the types of cleaners you can use on them.

In the main areas of the typical home — the living areas, kitchen, bathroom, and laundry — there are some jobs that will require specific cleaning solutions. As you read through this section, you'll find formulas, blends and treatments suitable for the different areas.

Use the recipes given here, and feel free to experiment on your own. Your choice of essential oils to perfume your own products will give you great freedom in terms of scent. You will also find that, with experience, you will come to know what proportions of ingredients work best.

A word of warning: don't overdo the use of essential oils, thinking that if a little smells good then a whole lot more will smell much better. Remember that the oils are concentrated, and a few drops are all you will need to add scent. Those little bottles will last a very long time, especially if you have a range of them. Always store your bottles of essential oils in a cool, dark place.

As a general guide, most of the recipes in this book will keep in storage for up to two weeks, however using fresh blends is better.

Now, let's start on that walk. We'll begin in the kitchen, often the hub of the home.

NOTE

You can adjust all the recipes that follow by either increasing or decreasing ingredients to suit your particular needs. As a guide, here is what the main ingredients will do:

- Vinegar *cuts grease, and is a mild disinfectant, bleach and deodorant.*
- Lemon juice *cuts grease, and is a mild bleach and deodorant.*
- Baking soda *removes stains and deodorizes, and is a mild abrasive.*
- Borax *removes stains and deodorizes, bleaches and disinfects.*
- Salt *is a good scourer, and an antiseptic and disinfectant.*
- Eucalyptus oil *is an antiseptic, disinfectant and deodorant.*
- Essential oils *add scent and also power to your cleaning products.*

A SPARKLING KITCHEN

The kitchen is the place the household tends to congregate, and where people come and go, leaving in their wake all the debris associated with cooking, chopping and pouring — and a lot else too.

Baking soda and vinegar work wonders in the kitchen, as they do elsewhere. Keep baking soda in a shaker under the sink and sprinkle it onto any stains that appear. And keep a spray bottle of white vinegar in there as well for a quick swipe at greasy spots and the finishing touches.

Salt is great in the kitchen too, and not only for cooking. Use it to scour your chopping boards — it will clean and disinfect at the same time.

Countertops will sparkle if you simply rub them with baking soda and a damp cloth, then wipe over with a cloth dipped in white vinegar to remove any powdery residue.

It won't take long for the chrome rings on the stove top to become stained and greasy. But they'll come up looking shiny and new if you soak them for just a few minutes in a sink of hot water with 2 tablespoons of baking soda added. Give them a rub with a scrubber, rinse them well, and wipe with a dry cloth or towel before replacing them.

KITCHEN APPLIANCES

Use the *Zingy Appliance Blend* to remove grease and grime from the surfaces of appliances such as dishwashers, microwave ovens, freezers and refrigerators. Your appliances will zing with citrus freshness.

Zingy Appliance Blend

- 2 teaspoons borax
- ½ cup hot water
- ¼ cup white vinegar
- ½ cup lemon juice
- 2½ cups water
- 2 teaspoons liquid castile soap
- 6 drops sweet orange essential oil
- 6 drops lemon essential oil

Dissolve the borax in the hot water, then pour this and the other ingredients into a spray bottle, and shake well. Spray onto the surface of any kitchen appliance. Wipe the appliance over with a damp, clean cloth.

The insides of your appliances will also want attention. They can become grimy and in need of a good clean. Here are recipes for cleaning inside the freezer, the microwave oven and the oven.

Fabulous Freezer or Fridge Formula

Before using this in your freezer, either turn up the freezer temperature (making it warmer) for ten minutes, or leave the door open for the same amount of time. Remove any food and pack it in ice, either in the sink or in a cooler.

- ¼ cup baking soda
- 1 cup water
- 6 drops rosemary essential oil

Mix ingredients together. Dip a clean, soft cloth into the mixture, and wipe down the inside of the freezer. Then wipe with a cloth dipped in white vinegar. Finally, wipe with a dry cloth. Turn the temperature down again (to make cooler), and replace food.

Marvelous Microwave Cleaner

Before using this cleaner on the inside of your microwave, remove the glass turntable, and clean it separately with a wipe of white vinegar on a moist cloth.

- ½ cup baking soda
- 1 tablespoon white vinegar
- 4 drops lemon essential oil
- 4 drops thyme essential oil

Blend these ingredients into a paste, and use a soft, clean cloth to apply the paste to the interior of the microwave. Wipe over with a damp cloth and allow the microwave oven to dry naturally, with the door open, before replacing the turntable.

Amazing Oven Cleaner

One of the worst jobs in the kitchen has got to be cleaning the oven, especially when there is an accumulation of spills that have not been cleaned away. Here is a very effective oven cleaner. Seeing the results will make this particular job a pleasure!

HINT

To clean away fresh spills in the oven easily and effectively, pour salt over the spill immediately. Wait for a few hours, then scoop or sweep up before wiping down with a damp cloth.

- ½ cup salt
- ¼ cup borax
- ½ cup baking soda
- warm water

Mix together the dry ingredients, and add enough water to make a paste. Apply it liberally to the walls, ceiling and floor of the oven. Leave for at least an hour. Wipe down with a clean cloth wrung out in white vinegar (with a few drops of lemon essential oil added, if you wish). Wipe down once more with a clean, damp cloth.

The Good Garbage Cleaner

If you have a garbage can, you'll know how grubby it can become with the constant depositing of debris and leftovers. To both clean and disinfect, wipe your garbage can over with this solution:

- 1 tablespoon borax
- 3 tablespoons hot water
- 4 drops tea tree essential oil

Mix together. Wipe solution all over the outside and inside of the bin. Wipe down with a clean damp cloth.

Super Sink Cleaner

If it's left uncleaned, a stainless steel sink can lose its luster and look very dull indeed. Bring back its natural shine with this simple mix:

- 2 tablespoons baking soda
- 2 tablespoons salt

- 1/4 cup white vinegar

Sprinkle the baking soda and salt all over the sink, and give it a good scrub with a scrubber. Make sure to get into all those little grooves and into the drain hole. Then rinse with hot water and give a final wipe over with a cloth dipped in the white vinegar.

Sensational Citrus Dishwashing Liquid

Make your own dishwashing liquid using castile soap and the essential oils of your choice. This recipe is a particular favorite of mine. It will make enough for many sessions of washing the dishes. You'll love the freshness of the citrus fragrance, and it will linger for some time, especially if you end by wringing out the dishcloths in it.

- 12 fl oz (375 ml) liquid castile soap
- 15 drops lemon essential oil
- 8 drops sweet orange essential oil
- 8 drops lime essential oil

Mix all ingredients in a squirt bottle. Shake the bottle each time you use the mixture. You will need only a couple of tablespoons each time you wash the dishes.

The Dishwasher

It's best to look out for commercial organic brands here. There are a several good biodegradable products available from supermarkets and health-food stores.

A GLISTENING BATHROOM

You can use organic products in many ways in the bathroom. Choose a product that works best for you, or try out a different product each week for a bit of variety.

- White vinegar does a good basic job in the bathroom. Apply it on a cloth. You could also warm the vinegar slightly, pour it into a spray bottle, and spray generously over all surfaces to be cleaned. Leave for a while, and then scrub if necessary before rinsing well.
- Baking soda will clean up surfaces that are not too badly stained. Pour a small mound into the bath or sink, and dip a moist cloth into it to clean away scum buildup. For slightly tougher stains, apply the soda with fine steel wool.
- Borax is also an effective bathroom cleaner: use it on a cloth or scourer to clean bathtub, sink and shower.
- A dark-colored bathtub is best treated with eucalyptus oil. Wring out a cloth in hot water, add a few drops of the oil, and rub over the tub to clean up stains and leave a shining surface.
- Chrome taps will shine like new if you simply wipe them over with a damp cloth dipped in white vinegar. Then polish with a dry cloth.
- To remove both mildew stains and soap buildup from the shower curtain, scrub with warm white vinegar. Then wash in warm soapy water, rinse, and hang out to dry in the fresh

air. You can help prevent the growth of mildew on shower curtains by soaking them in a tub of water to which you've added 1 cup of salt.

- Glass shower doors will come clean of any soap buildup if you rub them down with a mixture of salt and lemon juice, or borax and lemon juice. Rinse clean.
- Tiles will come clean of soap scum with a scrub of baking soda. Then wipe down with white vinegar to clean away powdery residue.
- To clean grouting between bathroom tiles, use a toothbrush dipped in baking soda.

The bathroom, perhaps more than most rooms in the house, can become very grubby in corners and along edges. What can you do about those stubborn or longstanding stains on the bath, sink and shower? Try any one of the following recipes.

Fabulous Bathroom Fizz

You'll love the bubble and fizz of this remarkable cleaning blend. It certainly looks as though it's getting at the grime!

- ½ cup white vinegar
- ¼ cup baking soda

Mix together the vinegar and baking soda, and use the mix to scrub the affected area with a cloth or brush. Leave it for 30 minutes to do its work before giving another scrub. End up by rinsing the area with warm water.

Super-Tough Scouring Powder

Not only does this powder clean dirty surfaces, but it disinfects as well.

- ½ cup borax
- ½ cup baking soda

Mix together, and keep in a shaker under the sink. Use on all surfaces, and brush into grouting to remove dark staining. Once it's applied, leave for half an hour before rinsing clean.

The Lemon Chaser

The lemon scent of this cleaner will linger in the air, leaving a fresh outdoor fragrance. The lemon juice will have a bleaching effect, enhancing the work of the borax in attacking stains.

- ½ cup borax
- ¼ cup lemon juice
- 6 drops lemon essential oil

Mix ingredients together to a paste and apply to soiled area. Leave for 30 minutes or more, then rinse well.

Magic Mirror Wipe

After a shower or bath, it can be impossible to see anything in the bathroom mirror. This recipe will ensure a fog-free mirror, and a shiny one at that!

- 1 cup white vinegar
- 5 drops eucalyptus oil

Combine ingredients in a spray bottle. Spray, and then use a dry cloth to wipe down.

THE TOILET

Forget about bleaches and artificially scented pine cleaners. For a quick clean, use the *Fabulous Bathroom Fizz* formula on page 41, or try the following recipes.

General Antigerm Cleaner

Here is a cleaner that can be used anywhere, but is especially effective around the toilet area, where germs love to breed.

- 2 cups water
- 1/4 cup liquid castile soap
- 1 teaspoon eucalyptus or tea tree essential oil

Mix these ingredients together in a spray bottle, shaking well. Use this cleaner to clean all toilet surfaces, then wipe over with a damp cloth.

Toilet Bowl Sweetener

A pleasantly fragranced cleaner to remove all traces of dark rings around the toilet bowl.

- 1/2 cup white vinegar
- 1/2 cup borax
- 1 tablespoon liquid castile soap
- 3 drops sandalwood essential oil
- 1 teaspoon vanilla essence

Combine the ingredients and pour them into the toilet bowl. Leave for several hours or overnight, give a good scrub with the toilet brush, and flush clean.

GLEAMING LIVING AREAS

Most living areas get a lot of use. Household members and visitors go back and forth, bringing with them the usual dust and grime that is part of everyday living both indoors and outdoors. When it's time to do spring cleaning, you will need to restore and revive couches and other furniture, wall hangings and various treasures. Here are some are ideas for organic cleaners you can use at this time. Make up large quantities if you are embarking on a total spring cleaning!

If, however, you prefer to keep such big efforts to a minimum, you will need to do ongoing maintenance work instead. Do so on a weekly or monthly basis, so that nothing becomes too big a job. This is especially true for stains, which respond far better to treatment if they are tackled earlier rather than later. You could check the upholstery on your living room furniture once a month or so, and do any spot cleaning then rather than waiting until the whole piece needs a thorough clean.

The furniture in living areas will differ markedly from one home to another. Here are some general ideas for cleaning these areas, and the items in them. Refer to the "A–Z" guide for further items. Note that flooring is dealt with separately, in the next section.

As you become more accustomed to making up cleaning products yourself from raw ingredients, you will come to know what all the individual products can do, and what you can achieve with combinations of them. Soon you will be able to come up with your own variations on the recipes that follow here.

CLEANING WALLS

The walls of the living areas (and in fact of all areas in the home) will benefit from an occasional dusting. Use a soft cloth on a long-handled broom or brush, or use the brush attachment of your vacuum cleaner to clear dust and any cobwebs that appear.

For finger marks around light switches and door handles, use a damp cloth dipped in white vinegar. If the marks are quite grimy, add a few drops of eucalyptus oil to the cloth. For marks elsewhere on walls, try using a gum eraser, (available from newsagents or artist supply stores). This may be all that's necessary.

You may decide to wash down your walls instead. If you want to transform their dingy surfaces, use the *Wonderful Wall Wash* — the recipe follows. But first, a few hints for cleaning walls:

1. It may seem illogical at first, but you should always clean walls from bottom to top. As you clean, the washing water will dribble down the wall along with the grime from the top, and the grime will settle farther down the wall, leaving streaks that will be difficult, if not impossible, to remove.
2. Use plastic covering on the floor if you want to keep the flooring from becoming wet, but cover the plastic in turn with some old fabric — perhaps an old sheet — to avoid slipping.
3. You'll find that, because you are holding your arms up as you clean, dirty water will be running down your sleeves. Avoid this by attaching toweling or other cloth to your wrists with rubber bands, and wear an old shirt.

Wonderful Wall Wash

The eucalyptus oil in this recipe acts as a cleaning agent, and adds its properties as an antifungal and antibacterial wash.

- 4 cups warm water
- 2 cups white vinegar
- 20 drops eucalyptus or tea tree essential oil

Combine ingredients in a bucket. Use a sponge or clean cloth to wipe down the walls, beginning from the bottom and working your way to the top. Have a second bucket filled with warm water only, and use a second sponge or cloth to wipe down the walls after applying the mixture.

CLEANING WALLPAPER

Try to keep your papered walls dusted regularly so there's less need to wash them down. If your wallpaper is nonwashable, clean away marks by rubbing in a little borax powder and then brushing away. A gum eraser will remove many sorts of marks, as will a piece of white bread squeezed into a tight ball. Rub gently with either of these.

The fabric part of lampshades can be treated in the same way. They will come clean if you sprinkle on a little baking soda or borax, leave for a few seconds, and then use a soft brush to dust the powder away.

Washable Wallpaper Cleaner

- 1 teaspoon liquid castile soap
- 2 cups warm water

Combine ingredients. Wipe the wallpaper in a circular motion, covering one small area at a time. Then wipe over with a damp, clean cloth. Avoid saturating the paper when washing it down.

FURNITURE

Look after your timber furniture by making sure that it is not exposed to strong sunlight or extremes of temperature.

Some timber furniture will benefit from an occasional polish, but this will not suit all types. If your furniture has a lacquered finish, for example, then adding an oily mixture may only serve to attract more dust. In this case, you will only need a simple wipe with a damp cloth sprinkled with a favorite essential oil.

Timber that is unfinished or old will respond to *Perfect Furniture Polish* (see recipe p48). Rub it onto the underside of tables and other surfaces too, to prevent cracking. Do not use on furniture with a French polish finish.

For scratches in timber furniture, rub in lemon juice and vegetable oil, mixed in equal quantities. The white marks left by hot cups can be removed by rubbing in a mixture of equal amounts of olive oil and salt. You could also rub mayonnaise into the marks, and leave overnight. When you wipe away the mayonnaise the next day, the marks should be gone.

Perfect Furniture Polish

Apply this polish to your timber furniture after dusting it clean. It will feed and condition the wood.

- ½ cup olive oil
- 1 tablespoon white vinegar
- 4 drops rose essential oil (or oil of your choice)

Combine these ingredients in a screw-top container. Shake well. Use a soft, clean cloth to rub in the polish.

The Clean Cane Formula — for cane, bamboo, rattan and wicker

Clean pieces made from these materials regularly, using a small, stiff brush. To deal with grimy corners and other dirty areas, use a toothbrush dipped in soapy water. Then wipe down with toweling.

Sometimes this type of furniture will yellow as it ages. To help counter yellowing, wipe down with salty water every few months.

Use the *Clean Cane Formula* on your cane, bamboo, rattan or wicker pieces. The lavender will give a sweet, old-fashioned scent. Substitute another oil if you prefer.

- ¼ cup liquid castile soap
- 1 cup water
- 6 drops lavender essential oil

Combine these ingredients in a spray bottle. Spray onto the piece, leave for a few seconds, and then wipe down with a damp, clean cloth.

LAMINATED AND PLASTIC FURNITURE

The following recipe will clean and revive these surfaces beautifully.

All-purpose Cleaner

Substitute lemon juice for white vinegar if you wish, and use sweet orange or lemon essential oil for a delightful citrus fragrance.

- 1 tablespoon borax
- 2 cups very hot water
- 1 tablespoon white vinegar
- 1 teaspoon liquid castile soap
- 10 drops essential oil of your choice

Dissolve the borax in the water, and then add the other ingredients. Use a spray bottle to apply the cleaner to laminated and plastic surfaces. Wipe clean with a damp cloth.

UPHOLSTERY

Most upholstery will benefit from regular vacuuming and, if possible, shaking out. Use the appropriate nozzle attachment of your vacuum cleaner to get at those corners and crevices where dirt collects. Take care, though — some fabrics are more fragile than others, and need a lighter approach. Suede is an example: you can use a vacuum nozzle to run along seams, but the fabric should just be brushed gently, with a soft brush.

The first — and most important — rule of stain removal is essential for upholstery: treat the spill immediately to minimize the possibility of staining.

If you can, store cloths and toweling in various places around your home so you can access them quickly in case of an accidental spill.

NOTE

- *However safe a spot remover may be, it is wise to test it on a small patch first. Always consider whether it is worth attacking a stain if there is a possibility that the treatment may cause more problems. Dab gently, and start from the outer corners of the stain so you don't spread it.*
- *Commercial nonorganic spot removers contain a toxic chemical that's also used for dry cleaning. Avoid these, and use your own homemade spot remover. See the recipe below.*

Super Spot Remover

This spot remover is effective on most fabrics. It can be used for various types of stains, such as chocolate, coffee, mildew, blood and urine.

- 1 tablespoon borax
- 1 cup hot water
- 6 drops eucalyptus oil

Dissolve the borax in the water, allow to cool, then add the oil. Pour the mixture into a jar or a spray bottle, depending on how you want to use it. Then either sponge or spray the mixture onto the stain, allow it to dry, and sponge again with a damp cloth. Reapply if necessary.

Gentle Leather Blend

This blend is equally good for cleaning dirty marks from leather or from vinyl upholstery. Follow it up with the conditioning treatment.

- 1 tablespoon liquid castile soap
- 2 tablespoons white vinegar
- 1 cup warm water

Combine these ingredients and apply with a soft cloth. Wipe over again with a clean, damp cloth, then finish with a soft, dry cloth.

Conditioning Treatment for Leather

- ½ cup white vinegar
- ½ cup olive oil
- 4 drops lemongrass essential oil

Mix together in a spray bottle, spray onto the leather and polish with a soft cloth.

CURTAINS

If you regularly dust down your curtains with a soft brush or a vacuum, then you will not have to do as much serious cleaning. Ingrained dust can leave streaks that cause bad stains over time.

With curtains that are washable, remove all hooks or other attachments, and soak in the laundry tub in cool water for an hour or so. This will help soften the fabric and allow the dust and grime to surface. Then either hand wash or gentle machine wash the curtains, using liquid castile soap. Add a few drops of your favorite essential oil if you wish. Dry outdoors, but not in direct sunlight.

WINDOWS

Washing windows can seem a huge task, especially if you have many of them in your home. Make the job as easy as you can by not cleaning windows when the sun is beating down through them, and for the best result do the outside and the inside at the same time. That way, you will get to have a lovely clear view and will be able to enjoy the results of your good work at once.

Unlike wall cleaning, with window cleaning you should start from the top and work your way down. Avoid using too much liquid, or you'll have a lot to mop up later. Have a window squeegee handy if possible, although you can do just as good a job without one. A chamois is good for a streak-free finish. Have a few extra cloths ready too, lint-free to avoid having bits of lint trailing across the window at the end.

See-through Window Cleaner

Use the following formula for a grubby window. However, if windows are particularly grimy, add a tablespoon of liquid castile soap for extra power to cut through the dirt.

- 8 cups hot water
- 1 cup white vinegar
- 8 drops peppermint essential oil

Combine ingredients in a bucket and use a sponge to apply to windows, starting from the top. Wipe away with a clean cloth or the squeegee blade, then wipe over again with a clean damp cloth. Remove last traces of dampness with a chamois.

FRESH FLOORING

Whatever type of flooring you have in your home — and most homes have several different kinds — floors come in for a tough time. Dirt and mud, things dropped and spilled, dragged furniture, and the constant movement of people and animals all add wear and tear. Little wonder, then, that floors need to be cleaned frequently.

CARPET

For good health all round, keep your carpet vacuumed regularly. And if you have pets who make the carpet their home, vacuum more often. Fleas love both the warmth of the carpet, and the opportunities it offers to hide away.

Carpet is one of the first parts of the home to look worn and grubby, especially with little shoes and feet and fingers on it. Patterned carpet is a boon in a home with children. If your carpet shows every mark, however, be particularly vigilant about spills. Attend to them immediately. Keep old towels handy so that whatever falls onto the carpet can be quickly "stepped on" with toweling, which will absorb more than the average cloth. Throw the towel down and literally step all over it, adding more towels to suit the extent of the spill.

Keep small bottles of soda water handy in the cleaning cupboard to treat spills as soon as you can. Spray or pour the soda water over the area, and the fizzing action will bubble the staining material to the surface, allowing you to towel it away. If the stain persists, use *Super Spot Remover* (for recipe see p50) for upholstery, or refer to the "A–Z" section for the specific stain removal method.

Here is a recipe that will lift the grime from your carpet and leave the room smelling fresh and appealing. Use it if you feel that your carpet needs a lift, and vacuum well before you begin.

Lavender-fresh Carpet Cleaner

These quantities are enough for an average-sized room. Remember to test on a small, inconspicuous area first, to make sure that your carpet will respond well.

- 1 tablespoon liquid castile soap
- 2 cups warm water
- 1/4 cup white vinegar
- 6 drops lavender essential oil
- 3 drops clary sage essential oil

Add the castile soap to the water, swish around, and then add the other ingredients. Use a sponge to apply the mixture to the carpet, covering one small area at a time, and then use a brush to rub it gently into the fibers. Don't use too much of the cleaner at a time — the carpet should be dampened, not saturated. Allow to dry, then vacuum well.

TIMBER FLOORING

Like timber furniture, flooring with a protective finish will not benefit from a polish. The following formulas will clean a timber floor effectively and leave a pleasing fragrance.

Simple Timber Floor Wash

- 2 cups cool tea — this could be a herb tea, such as chamomile or peppermint

Wring out a mop in this solution and wipe over your polished floorboards. The dust will come away, and the floor will smell and look fresh.

Exotic Timber Floor Wash

Bergamot and sweet orange are a lovely combination, but you may prefer your own combination of oils.

- 4 cups hot water
- 4 cups white vinegar
- 10 drops bergamot essential oil
- 10 drops sweet orange essential oil

Combine all ingredients in a bucket and mop or sponge the floor, covering a section at a time. Avoid overwetting the floor. Wipe over with a dry mop or cloth.

TILE, VINYL, CORK, LINOLEUM, SLATE FLOORING

For simple week-to-week cleaning, use the following recipe, which will brighten up the flooring everywhere in your home.

Basic Floor Wash

- 3 cups white vinegar
- 12 cups (3 quarts) hot water
- 20 drops pine essential oil

Mix together in a bucket, and sponge or mop across the flooring, taking care not to splash too much liquid around.

Tougher Floor Wash

Use this recipe when flooring is dirty and greasy and you need more cleaning power.

- 12 cups (3 quarts/3 liters) hot water
- 3/4 cup liquid castile soap
- 1 1/2 cups white vinegar
- 20 drops tea tree essential oil

Mix ingredients together in a bucket, and mop or sponge in small sections at a time, rubbing sufficiently to dislodge the dirt and grease. If you like, you can go over the floor with a damp mop or sponge later, but this is not necessary.

STAIN-FREE LAUNDRY

Let's start with a few guidelines for laundry work.

1. Sort the clothes, and wash whites together. By following this simple rule you can avoid the dingy look white clothes can get when they pick up the dye — even in microscopic amounts — from darker clothing.
2. Presoak or pretreat clothing and other items that are heavily soiled.
3. Empty pockets before anything goes into the wash. Facial tissues and other bits and pieces hiding in pockets will cause havoc with your washing, and darker washing in particular.

4. For their well-being, close zippers. However, you'll need to undo buttons for more effective cleaning.
5. Floating strings (as on aprons) or ties (as in men's pajama bottoms) should be tied in a bow to prevent tangling.
6. Clothing of any fabric that tends to pill, such as jersey knit, should be turned inside out.
7. Any item in a color that might bleed should be washed either separately, or with similarly colored items.
8. Use white vinegar in the final rinse to remove all signs of soap. Add borax to the final rinse as a fabric softener.

Water temperature can be a dilemma when it comes to doing the laundry. Using cold water may save on energy, but will it do a good job in every case? The answer is that different fabrics will get cleanest in different water temperatures. Let this chart guide you.

Temperature	Best for
Use cold water	• Brightly colored items, to stop dyes from bleeding • Any item that might shrink in warmer temperatures, such as some cottons • All rinsing
Use warm water	• Permanent-press clothes • Wool that is washable • Clothing in synthetic fabrics • Dark-colored items • Most cottons (for 100 percent cottons, use lukewarm water so they won't shrink)
Use hot water	• Items that are very soiled • Whites, to help keep them white • Towels • Sheets

PRETREATING

Sometimes clothing and other items will need attention before they go into the washing machine — a dab of stain remover, or a good soak. If you're soaking laundry, use either the tub or a bucket, and make sure there is enough room for the amount you are soaking. It is counterproductive to bunch together too many items requiring treatment.

To remove stains from fabric, mix 2 teaspoons of cream of tartar with enough water to make a paste. Apply liberally to cover the stain, leave to dry and then wash as usual.

General Pretreatment Spray

When you are about to wash the clothing, use this formula to give heavier stains a quick treatment:

- 1 tablespoon liquid castile soap
- 1 tablespoon glycerin
- 1 tablespoon borax
- 1 cup hot water
- 6 drops tea tree essential oil

Dissolve the borax in the hot water, then add the other ingredients. Use a spray bottle to apply this mixture to whatever stains you find on laundry items before washing. Leave the mix on for 2 minutes to start doing its work, and then wash the items as usual.

Presoak blend

- ½ cup borax
- 3 cups hot water
- 1 cup white vinegar
- 6 drops eucalyptus oil

Dissolve the borax in the hot water, allow to cool, and then add the other ingredients. Before washing as usual, soak items in this mixture for 2 hours, either in the laundry tub or in a bucket. Very tough stains may require longer soaking, or an application of glycerin to soften and release the stain before the usual laundering.

WASHING LIQUID/LAUNDRY LIQUID

The quantities in the following formula for *General Use Laundry Liquid* will give you enough for between 8 and 12 loads of washing, depending on how much you are washing and how soiled the washing is.

Use one of the containers from the commercial laundry liquids you used to buy to make your own safe and effective laundry liquid. Wash the container well before using it.

General Use Laundry Liquid

You can vary the essential oil used in the recipe: use tea tree for its antibacterial and antifungal properties, or sweet orange to help remove stains, or a combination of both.

- 1 cup washing soda
- 2 cups hot water
- 1/4 cup liquid castile soap
- 1 cup baking soda
- 8 drops tea tree or sweet orange essential oil

Begin by dissolving the washing soda in the water. Shake the container to help it dissolve. Then add the other ingredients. Use up to 1/2 cup per wash load.

Three-Cup Washing Powder

This powder is called *"Three Cup"* because you need just a cup each of its main ingredients. The sandalwood gives a woody scent that is not too sweet.

- 1 cup washing soda
- 1 cup baking soda
- 1 cup pure soap flakes (you can finely grate your own bar of pure soap)
- 6 drops sandalwood essential oil

Combine all ingredients and use up to 1/2 cup per wash load, depending on how soiled and how large the load is.

Safe Bleaching Solution

The lemon juice in this recipe means you'll have to keep it in the fridge when it's not in use. If you prefer, however, you can halve the quantities, and make up a fresh batch each time you need it.

- ½ cup lemon juice
- 2 cups water
- 8 drops lemon essential oil
- ¼ cup borax

Mix together the lemon juice, water and lemon oil, and soak laundry you want to whiten in this solution for at least an hour. Then wash as usual, adding the borax to the rinse cycle. For extra natural bleaching power, hang the clothes out to dry in the sunshine.

Gentle Wash for Delicates

Use this recipe for those items that need a gentler touch, including baby clothes. The lavender and rose give a lovely Old World fragrance, but you can also experiment to find a blend that suits you.

- ¼ cup liquid castile soap
- 2 drops lavender essential oil
- 2 drops rose essential oil
- 3 cups warm water

Combine in a bucket or laundry tub. Wash your delicate items by gently squeezing them in this wash. Rinse with cool water.

CITRUS FABRIC SOFTENER

Many of the commercial fabric softeners have fragrance added, but none are as fresh and natural as this one. Make it in bulk in a large container from which it is easy to pour, and add to the final rinse for all your laundry.

- ½ cup borax
- 1 cup hot water
- 4 cups white vinegar
- 8 drops lemon essential oil
- 8 drops sweet orange essential oil

Dissolve the borax in the hot water, cool slightly, and add the other ingredients. Add a good splash of the mixture to the final rinse cycle.

The A–Z of stain removal and cleaning

This "A–Z" has over 200 entries of simple and effective treatments for cleaning and stain removal using only organic products.

It is intended to be a quick and easy reference guide, because when there's a spill, you'll want to attend to it as quickly as possible — not be hunting through a large volume while a bright-colored liquid makes its way ever more deeply into your carpet!

You'll find that this "A–Z" covers just about any stain you'll come across in the home. There are three types of entries here:

- ✸ Items that are *frequently stained*,
- ✺ Things that *cause stains*,
- ☛ *Alternative methods* of cleaning.

In some cases you'll be referred to formulas in the "Recipes: Formulas, blends and treatments" section on pages 33 to 61. It's a good idea to read through these recipes when you have time, rather than when you're facing a fresh stain.

Many of these formulas and blends have essential oils added to them, both for their added power and their pleasing perfume. The suggested treatments in the "A–Z" do not include essential oils in their formulas, but you can add your own. If you intend to do so, read the section on essential oils on pages 23 and 24.

Using organic products is not only healthier for people and the environment, but will also make cleaning more interesting for you. You can experiment with ordinary household products, and discover new uses for them. With your new knowledge of their effectiveness, there's no limit to the variations and combinations you can come up with as you put these wonderful products to good use in your home.

PLEASE NOTE: All care should be taken when using the following suggested stain-removal methods on fabrics, carpets and delicate paint finishes. In these cases you should always test the suggested method on a small, inconspicuous area before using.

ACID

Mop up the spill quickly and then soak the area in cold water. To neutralize the acid, apply a paste of baking soda and water. When the paste is dry, brush away before sponging down with water.

ADHESIVE LABELS

These will come away easily if you apply a little eucalyptus or tea tree essential oil, and rub gently with a cloth. You could also try peanut butter — this loosens some labels effectively.

ALCOHOL

There are many different types of alcohol, and alcohol stains will differ in color and degree of difficulty when removing. For any alcohol spill, however, the most effective course of action is a quick application of soda water. Treat all surfaces with alcohol stains in the same way. Sprinkle, spray or pour soda water onto the affected area.

Next, mop up the liquid with a towel. If carpet is stained, stepping on the towel or paper will help the liquid to be well absorbed. Repeat the process until all signs of the alcohol are removed. If alcohol is spilled on your clothing while you are eating out, ask for soda water, and use napkins to mop up the excess liquid. Here are some alternative methods for removing alcohol stains.

✸ Fabric

Soak the item in a bucket of cold water, with ½ cup of white vinegar added to it. After soaking for a couple of hours, wash as usual. For a stubborn stain, soak in a weak borax solution: 1 tablespoon of borax dissolved in a ½ cup of hot water and added to a bucket of cold water.

✸ Carpet

Use a towel to soak up the spilled liquid, and then sprinkle the area well with either baking soda or salt. This will absorb the fluid remaining in the carpet. Once it is completely dried (which could take up to several hours), vacuum thoroughly.

☛ Sponge the area with white vinegar, especially if the alcohol has already stained the carpet.

✸ Furniture

Act quickly to soak up the spilled liquid with an old towel. Then, in the case of timber furniture, a little olive oil rubbed into the grain will revive it. For other furniture, add a few drops of your favorite essential oil to a damp cloth, and wipe down.

See also **Red wine**

ALUMINUM

Aluminum saucepans will come clean of stains if you drop in a few lemon slices, with some water, and bring to a boil.

ANTIPERSPIRANT

See **Deodorant**

B

BABY FORMULA

Soak the affected area in white vinegar before washing in lukewarm water.

BABY VOMIT

While stain is fresh, wipe away excess and run cold water through fabric to remove all traces.

If it's not possible to do this, for example if you are unable to change, then dab at the area with soda water.

If the stain has dried, soak the affected area in white vinegar before washing as usual.

See also **Vomit**

BAKING PANS

To remove the burned bits stuck to the bottom of a baking pan, apply a liberal amount of paste made from baking soda and water. Leave overnight, discard what has come away, and then wash the pan in hot, soapy water.

BAKING SHEETS

See **Cake pans**

BALL-POINT PEN

Fabric

Back the marked area with a rolled-up towel or cloth, and dab at the stain with eucalyptus oil until it comes away. Please note: avoid using this method on delicate fabrics, such as silk. Test on a small, inconspicuous area of the fabric before using.

Carpet

Using a cloth dipped in a mixture of equal parts of milk and white vinegar, dab at the ink mark until it disappears.

See also **Ink**

BANANA

See **Fruit**

BARBECUE COOKING GRATE

See **Cooking grate**

BARBECUE SAUCE

Fabric

Flush immediately with cold running water, then soak in a solution of ½ cup of white vinegar and ½ bucket of water for half an hour, before washing as usual.

If stain persists, apply a paste of cream of tartar and water. Let the paste dry, then wash as usual.

Carpet

Scrape away any excess, then spray with soda water. Press toweling into the area to absorb as much liquid as possible. Repeat the process if necessary to remove all traces of the stain.

BATTERY ACID

See **Acid**

BEER

See **Alcohol**

BEETS

Fabric

If possible, run cold water through the stain to remove as much of it as possible, or soak the item in milk for a few hours before washing as usual.

For a more stubborn stain, apply a paste of cream of tartar and water, let it dry, and then wash as usual. You could also use the *General Pretreatment Spray* (for recipe see p58).

Carpet

Sprinkle soda water over the stain and blot up immediately with toweling, stepping onto it to absorb as much of the liquid as possible. If stain persists, apply *Super Spot Remover* (for recipe see p50).

BERRIES

Fabric

Clean up the spill and pour soda water onto the area, dabbing with a cloth to absorb the liquid.

☛ Run cold water through the stain to remove as much stain as possible. Then do one of the following:

- Dab the affected area with lemon juice, or half a lemon.
- Apply a paste of cream of tartar and water and allow to dry.

Follow up either treatment by washing the garment as usual, using lukewarm water.

If the stain is old, apply either cream of tartar paste or a little glycerin, and leave for a couple of hours before washing as usual.

✸ Carpet

Sprinkle soda water over the stain and blot up immediately with toweling, stepping onto it to absorb as much of the liquid as possible. Repeat the process until all traces of the stain have disappeared.

BIRD DROPPINGS

For the best results, let the droppings dry before scraping them away. Then sponge the stained area with a solution of 2 tablespoons of white vinegar and ½ cup of water.

BLOOD

NOTE

Do not use hot water on a bloodstain — it will set.

Fabric

If possible, run cold water through the stained area, then soak in cold water for several hours before washing as usual. Use lukewarm water.

In the case of a very small bloodstain, saliva will break it down. Spit on the stain, wait for a few minutes, then wash the garment as usual.

If the blood has dried on the fabric, apply a paste of borax and water. Once the paste has dried, brush it away, then wash the item as usual.

Carpet

For fresh stains, sponge the area immediately with cold water. If this is not enough to treat it, pour soda water over the stained area and blot quickly with a towel, stepping on it to absorb as much liquid as possible. Repeat the process until all traces of the stain have disappeared.

If the stain is dry, apply a paste of borax and water to the affected area, wait for it to dry, and then brush or vacuum away.

☛ Eucalyptus oil is also effective for treating bloodstains. Drop a little onto a wet cloth, and dab at the stain until removed.

NOTE

Avoid using eucalyptus oil on delicate fabrics, such as silk. Test by dabbing on a small, inconspicuous area of the fabric or carpet before using.

✷ BONE CUTLERY HANDLES

These will sometimes stain with age. To remove staining, rub gently with a cut lemon dipped in salt, then rinse the cutlery and dry thoroughly.

✷ BOOKS

Remove mildew stains by dabbing gently with a weak vinegar solution: 1 tablespoon of white vinegar to ½ cup of water.

Dust mildewed areas on books with cornstarch, leave out to air for a day or two, and then dust the powder away before replacing the books.

NOTE

To help prevent mildew affecting books, allow enough room around them for air to circulate.

✷ BRASS

For general upkeep, wash brass pieces occasionally in warm water, with a little liquid castile soap added.

To remove tarnish from brass, try this method: sprinkle salt onto half a lemon, and rub the metal. Then rinse thoroughly. Bottled lemon juice mixed with salt will also be effective.

Marks of corrosion that sometimes appear on brass will disappear if you rub them with a cloth dipped in buttermilk. Then rinse with water, and wipe dry with a soft, clean cloth.

BRASS CURTAIN RINGS

To clean up brass curtain rings, soak them in a warm solution of ½ cup of white vinegar and ½ cup of water for half an hour. Then rinse, and dry thoroughly.

BRICK

The bricks around a fireplace can become very stained over time. To remove staining, wash with hot white vinegar before wiping down with a clean, wet cloth.

BUBBLE GUM

Fabric

Place the affected item in a plastic bag, and leave in the freezer for a couple of hours. Then use a knife to scrape the gum gently off the fabric. To get the last fragments of gum out, either soak the garment in white vinegar or dab with eucalyptus oil. Then wash the garment.

Carpet

Add eucalyptus oil to a cloth and dab at the area. Then scrape away the gum with a knife.

NOTE

Avoid using eucalyptus oil on delicate fabrics. Test by dabbing on a small, inconspicuous area of the fabric or carpet before using.

BURN MARKS

Fabric

If the fabric is white, rub lemon juice into the mark, and leave the item to dry in the sun.

☛ Soak in a solution of 2 tablespoons of borax dissolved in a little hot water and then added to 2 cups of water. Soak until the mark fades, and then wash as usual.

✸ Carpet

Use fine steel wool to scrape away the burned fiber, or nail scissors to clip it away. Then sponge the area with a little white vinegar.

✸ Furniture

The white marks left by hot cups can be removed by rubbing in a mixture of equal amounts of olive oil and salt.

☛ Rub mayonnaise into the white marks, and leave overnight. When you wipe away the mayonnaise the next day, the marks should be gone.

✸ BURNED SAUCEPAN

For burned food stuck to the bottom of your pan, add one of the following:

- baking soda and water, mixed into a paste.
- baking soda and lemon juice, mixed into a paste.

Leave to soak, preferably overnight, and rinse clean.

☛ Add 1/4 cup of white vinegar and 1 tablespoon of baking soda. Bring to a boil and then leave to cool. Drain off the residue and then rinse clean.

✸ BUTTER

See **Grease**

C

✸ CAKE PANS

To remove those dark stains that sometimes appear on cake pans, rub with a wet cloth dipped in borax. Then rinse the pans thoroughly before drying.

✸ CANDLE WAX

See **Wax**

✸ CANE

Stained white cane will come up sparkling clean if you rub down with a wet cloth dipped in baking soda. Then wipe with a clean wet cloth to remove any powdery residue.

☛ You could also try the *Clean Cane Formula* (for recipe see p48).

✸ CARPET

For a particular stain, see specific entry.

For general information on removing stains from carpets and for a carpet-cleaning recipe, see pages 53 and 54.

✸ CEMENT

To clean stains from cement, scrub with a solution made from ¼ cup of borax dissolved in ½ cup of hot water and added to 2 cups of cold water. Then hose down.

To remove cement from clothing, soak in this mixture: 1 tablespoon of salt, 1 cup of white vinegar, and a bucket of water. Then wash as usual.

CERAMIC TILES

Clean away stains with a cloth dipped in white vinegar.

CHEESE SAUCE

See **Grease**

CHERRIES

See **Fruit**

CHEWING GUM

See **Bubble gum**

CHINA

A paste of baking soda and water rubbed into stains will help remove them without scratching the china.

CHOCOLATE

Fabric

Try soaking the stain in milk before washing as usual.

- Dab the stain with a cloth dipped in a mix of 1 tablespoon of white vinegar and ½ cup of water, then wash in lukewarm water.

A–Z

☛ Use *Super Spot Remover* (for recipe see p50).

✹ Carpet

Scrape away the excess with a knife, then spray with soda water. Press toweling into the area to absorb the liquid.

If a stain remains, sponge the area with *Super Spot Remover* (for recipe see p50).

✹ CHROME

To clean chrome pieces, simply wipe over with a soft cloth dipped in white vinegar.

Chrome will shine beautifully if you give it a rub with a piece of aluminum foil.

CHUTNEY

See **Barbecue sauce**

COCOA

See **Chocolate**

COLA DRINK

See **Soft drink**

COCONUT OIL

See **Grease**

COFFEE

✹ Fabric

Run cold water through the affected area and

then apply a paste of borax and water. Leave for an hour before washing as usual.

☛ Apply *Super Spot Remover* (for recipe see p50).

✷ Carpet

Spray soda water onto the spill and press toweling into it to absorb all the liquid. Repeat the process until stain disappears.

☛ Follow the first soda water application with *Super Spot Remover* (for recipe see p50).

✷ COFFEEPOT

Clean the inside of the pot by boiling up an equal mix of water and white vinegar. Let the mixture simmer for about ten minutes before rinsing well.

✷ COLLARS

To treat the ring of grime around collars, apply a paste of equal quantities of cream tartar and water. Leave to dry, then wash as usual.

✷ COMPUTER

All your office equipment will come up clean if you simply wipe it with a cloth dipped in white vinegar. Use a cloth that's lint-free for a totally clean surface.

☛ Add 6 drops of tea tree essential oil to $1\frac{2}{3}$ cups water, then spray onto a cloth before wiping equipment. Leaves a fresh fragrance.

✸ COOKING GRATE

While the grate is still hot after cooking, sprinkle it liberally with salt to absorb all the grease. Then, once it is dry, simply brush the salt deposit away.

If you have a cooking grate that needs cleaning before you can cook on it, give it a scrub with salt water. Use a soft brush, and wipe the surface dry with a cloth. Smear the surface with a little vegetable or linseed oil to prevent it from rusting.

✸ COPPER

If the copper piece has had lacquer applied to it, then it only needs dusting and the occasional wash in warm water. If the copper has become stained, however, try any of these methods:

- Use a mixture of lemon juice and salt. Simply sprinkle salt onto half a lemon and rub the metal. Then rinse thoroughly. Bottled lemon juice mixed with salt is also effective. Buff with a soft, clean cloth.
- Worcestershire sauce will also clean copper. Wipe it on, give it a good rub, and then rinse before polishing up with a soft, clean cloth.
- Here is an extremely good recipe for cleaning copper: mix together 1 cup of white vinegar, 1 cup of flour, and ½ cup of salt. Apply this thick paste to the copper piece and leave for a few hours. Rinse thoroughly, and wipe dry with a clean cloth.

COPPER COOKWARE

See **Copper**

CORRECTION FLUID

See **Liquid Paper**

COSMETICS

Fabric

Most cosmetic stains will come away if you dab at them with a solution of 1 tablespoon of borax dissolved in a little hot water and then added to ½ cup of water.

You could also apply a paste of cream of tartar and water. Leave to dry, then wash as usual.

Carpet

Use the same borax solution used for fabric.

See also **Lipstick; Mascara**

CRAYON

Rub gently with a damp cloth dipped in baking soda. Then wipe down with a clean cloth.

CREAM

See **Milk products**

CRYSTAL GLASSWARE

Wash with a solution of ½ cup of white vinegar and 1½ cups of hot water. Allow crystal glasses to dry naturally.

CURRY

Fabric

Dissolve ¼ cup of borax in ½ cup of hot water. Add to a bucket of water, and soak stained fabric before washing as usual.

Carpet

Spray soda water on the affected area, then press toweling into it to absorb all liquid. Then sponge with a solution of ½ cup of white vinegar and 1 cup of water.

☛ Apply *Super Spot Remover* (for recipe see p50).

CURTAINS

For stain removal treatment, see entries for specific curtaining materials — plastic, vinyl, fabric.

Curtains tend to become more easily stained over time if they are not brushed or vacuumed regularly. For washable curtains that need a good wash, remove all accessories and soak in a tub of cold water to help soften the fabric and allow the grime to come to the surface. Then machine wash gently using either liquid castile soap or *General Use Laundry Liquid* (for recipe see p59). Dry outdoors, out of direct sunlight.

CUTLERY

Avoid immersing any cutlery with handles of bone, wood or porcelain in water.

To remove stains from cutlery, rub with a damp cloth dipped in baking soda.

See also **Bone cutlery handles; Knives; Silverware**

D

DEODORANT

Deodorant can leave a large, dark stain on fabric in the area under the arms. Try any of the following methods before washing the garment as usual:

- Soak in white vinegar.
- Soak in a bucket of water to which has been added a solution of 1/4 cup of borax dissolved in hot water.
- Apply a paste of baking soda and water, and leave for a few hours.

DIAPERS

Use vinegar to disinfect and bleach diapers as well as deodorize them. Soak wet diapers in a solution of 1/4 cup of white vinegar mixed into a bucket of water, then wash in hot water.

With soiled diapers, empty feces into the toilet, rinse the diaper in cold water, and then soak for several hours in a bucket of warm water to which has been added a solution of 1/4 cup of borax dissolved in 1/2 cup of hot water. Wash as usual, and add 1/4 cup of white vinegar during the rinse cycle.

If possible, always dry diapers in the fresh air and sunshine.

DOORS

See **Painted doors**

DRAINS

To unblock drains, pour ½ cup of either washing soda or baking soda down the drain, followed by ½ cup of white vinegar. Leave it for an hour or so to do its work, then pour in ½ cup of salt and 5 or 6 cups of boiling water.

DRIPPING FAUCET

To remove the marks left by a dripping faucet, apply a paste of borax and water. Leave for an hour before rinsing. Repeat if necessary, leaving the paste on for longer the second time.

E

EGG

NOTE

Do not use hot water to remove egg stains: the heat will cook the egg and make it harder to remove.

Fabric

Wash away excess, then soak the item in a bucket of cold water with ¼ cup of salt added.

Carpet

Sponge with cold, salty water.

China and Cutlery

Wipe down with wet salt.

ENAMEL

Rub away the brownish marks that can form on enamel (such as refrigerator doors) with a cloth dipped in baking soda. Then wipe down with a cloth dipped in white vinegar.

☛ Use *Zingy Appliance Blend* (for recipe see p36).

EYE SHADOW

See **Cosmetics**

F

FABRIC

Please see relevant item causing stain.

FAT

See **Grease**

FECES

See **Diapers; Pet feces**

FELT

Small areas of stain will come away if you rub lightly with fine sandpaper.

FELT-TIP PENS AND MARKERS

See **Permanent markers**

FINGER MARKS

To remove finger marks from furniture and walls, wipe with a damp cloth dipped in any of the following:

- White vinegar
- Baking soda
- Eucalyptus oil

FISH STAINS

For clothing stained with fish slime, flush with cold water as soon as possible, then soak overnight in a solution of ½ cup of white vinegar in a bucket of cold water. Then, wash as usual.

FLOWERS

See **Grass**

FLY SPOTS

Those tiny fly spots will come away if you rub them with a damp cloth dipped in a little baking soda, then wipe down with white vinegar, applied on a damp cloth.

FOOD STAINS

Whatever the food stain, start by sponging with a solution of white vinegar and a little liquid soap. This may be enough to remove all signs of the stain on both fabric and carpet. If not, try any of the many other suggestions given either under entries for specific food items or in "Stain-free laundry," pages 56 to 61.

FRUIT

✷ Fabric

Sprinkle soda water over the stained area and blot up liquid. Then sponge with a solution of 1 tablespoon of borax and 1 cup of hot water, or soak in 1/4 cup of borax and a bucket of hot water, until the stain is removed. Wash as usual.

✷ Carpet

Sprinkle soda water over the stain and blot up immediately with toweling, stepping onto it to absorb as much of the liquid as possible. Repeat until all traces of the stain disappear.

☛ Apply *Super Spot Remover* (for recipe see p50) after one treatment with soda water.

See also **Berries**

FRUIT DRINK

See **Soft drink**

FRUIT JUICE

See **Fruit**

G

✷ GLASS

To clean stains from glass on areas such as oven doors, rub with a damp cloth dipped in baking soda.

To remove paint from glass, rub with hot white vinegar on a cloth.

For general glass cleaning, use a solution of equal parts of white vinegar and water. Spray onto the glass and wipe clean.

For windows, you can use *See-through Window Cleaner* (for recipe see p52).

✷ GLASS SHOWER DOORS

To clean away soap buildup, rub with a mixture of 2 tablespoons salt and ½ cup lemon juice, or 2 tablespoons borax and ½ cup lemon juice.

✷ GLASSWARE

See **Crystal glassware**

✷ GLUE

✷ Fabric

Warm 1 cup of white vinegar, and soak the affected area in it for about ten minutes. Rinse away the residue.

☛ Eucalyptus oil will effectively remove glue: drop a little onto a soft, clean cloth and dab the affected area until the glue comes away.

✱ Carpet

Soak a cloth in warm water and place it over the affected area, pressing the water into the glue. Leave the cloth there for an hour or so, after which the glue should be soft enough to come away. Repeat the process if necessary. Finish by rubbing a clean, damp cloth over the area to remove any last traces.

☛ Try dabbing at the area with a cloth onto which you've sprinkled a little eucalyptus oil or tea tree essential oil. If the glue has set and cannot be removed, use nail scissors to trim the affected area carefully away.

NOTE

Avoid using this method on delicate fabrics, such as silk. Test by dabbing on a small, inconspicuous area of the fabric or carpet before using.

GRASS

For grass stains on fabric, soak the stained area in white vinegar and then wash in hot water.

☛ Dab at the stain with eucalyptus oil before washing as usual.

GRAVY

Scrape away the excess, then treat as for **Grease.**

GREASE

Fabric

It may be enough to sprinkle the spot with flour or talc, and leave it for a while to absorb the grease. Then shake away the flour and wash the garment as usual.

☛ Run very hot water through the stain if possible, and apply eucalyptus oil to absorb the grease. Then wash as usual.

✸ Carpet

Sprinkle flour or talc over the stain and leave for several hours. Vacuum up, and then blot with a damp cloth sprinkled with eucalyptus oil.

NOTE

Avoid using eucalyptus oil on delicate fabrics, such as silk. Test by dabbing on a small, inconspicuous area of the fabric or carpet before using.

✸ GROUTING

Grouting will come clean when you brush moistened borax into it with a little, hard brush.

For stubborn stains, use *Super-Tough Scouring Powder* (for recipe see p42). Scrub into the grouting with hot water and leave for at least an hour before rinsing away.

GUM

See **Bubble gum**

H

HAIR DYE

✸ Fabric

Flush with cold water before soaking in a

solution of 1/4 cup of borax dissolved in 1/2 cup of hot water and added to a bucket of water.

Carpet

Act quickly: pour or spray soda water over the area and blot with a clean cloth or towel to absorb all the liquid. Sprinkle borax over the area, leave to dry, and then vacuum up.

HANDKERCHIEFS

Soak in a solution of 1/2 cup of salt and half a bucket of water before rinsing out under a running tap and washing as usual.

HANDS

To clean away stains left by fruit and vegetables, rub a cut lemon onto your hands, adding a little sugar for a slight abrasive action.

HEAT MARKS

See **Burn marks**

HEM MARKS

To eliminate those telltale hem marks when a skirt has been lengthened, dab white vinegar along the marks and iron (use a warm setting).

HONEY

Clothing

Flush with warm water to wash away residue. If a stain remains, apply a paste of borax and water. Leave to dry, and then wash as usual.

✸ Carpet

Sponge with warm water. If a stain remains, sprinkle with borax or baking soda. Leave to dry, then vacuum up.

I

ICE CREAM

See **Milk products**

INK

✸ Clothing

Try soaking the stained area in milk for a few hours. Then wash as usual.

☛ Dab vegetable oil onto the area so that the ink is thinned out, then blot it away with paper toweling. Keep this up until the stain begins to fade. Then wash as usual.

NOTE

Avoid using vegetable oil on delicate fabrics, such as silk. Test by dabbing on a small, inconspicuous area of the fabric before using.

✸ Carpet

Act quickly: blot up all you can, then spray or pour soda water over the area. Press toweling into the area to absorb as much of the liquid as possible. Sprinkle liberally with borax powder, leave to dry, and then vacuum up.

Hands

Rub vegetable oil onto stained fingers, wiping away the ink with paper toweling or cloth. Then wash your hands in warm water.

See **Ballpoint pen**

IRON

To clean the surface of an iron that is not Teflon-coated or nonstick, add 1 tablespoon of salt to 1/4 cup of hot white vinegar, and wipe this solution over the surface.

J

JAM

See **Honey**

JEWELERY

Warm soapy water (use liquid castile soap) will clean copper, diamonds, gold, jade, and mother-of-pearl. Wash them gently and quickly, and dry with a soft, clean cloth.

For opals and pearls, use a cloth dipped in glycerin before buffing up with a clean cloth.

To clean silver jewelery, use lemon juice. Either brush gently with an old toothbrush, orsoak for a while in lemon juice before rinsing in warm water and wiping dry.

JUICE

See **Fruit**

K

KETCHUP
See **Barbecue sauce**

KETTLE
See **Tean Kettle**

KNIVES
To remove stains from chef's knives, sprinkle salt on half a lemon and rub along the blade.

See also **Cutlery**

L

LACE
White lace can become yellowed with age. To remove the staining, soak the lace for a couple of hours in a solution of 4 cups of water and 1 tablespoon of borax. Dissolve the borax first in a little hot water.

Then wash with a mild soap, and rinse well. To stop the lace from pulling out of shape, pin it onto a towel that is lying flat. Then store in tissue paper to prevent discoloration.

LAMINATE
Stains can be removed from laminated surfaces by rubbing with baking soda on a damp cloth.

Finish with a wipe of white vinegar to clear away any deposit and leave the surface shining clean.

LAMP SHADES

To clean the fabric of a lamp shade, sprinkle with baking soda, leave for an hour, then dust away with a soft brush.

LEATHER

To clean leather, use *Gentle Leather Blend* (for recipe see p50).

LINENS

Linens can become yellowed with age. To restore their whiteness, sponge them with a solution of borax and lemon juice.

☛ Use half a lemon to press borax powder into the stained area, making sure that the area is well moistened. Leave to dry, then wash as usual. Dry in the sun.

LINOLEUM

To clean linoleum floors, use *Basic Floor Wash* (for recipe see p55).

LIPSTICK

Eucalyptus oil will remove lipstick from fabric or carpet. Simply add a few drops to a cloth and dab at the stain until it disappears.

NOTE

Avoid using eucalyptus oil on delicate fabrics, such as silk. Test by dabbing on a small, inconspicuous area of the fabric or carpet before using.

☛ Rub petroleum jelly or glycerin gently into the stain before washing.

LIQUID PAPER

Allow to dry, then scrape away the residue. Blot the area with a solution of 2 cups warm water, a teaspoon of liquid castile soap and a teaspoon of white vinegar. Then wash as usual.

M

MAKEUP

See **Cosmetics**

MANGO

See **Fruit**

MARBLE

For marble that has become stained, rub with half a lemon dipped in either salt or baking soda. Then finish off with a cloth wrung out in white vinegar.

MARKERS

See **Permanent markers**

MARMALADE

See **Honey**

MASCARA

To clean mascara marks from clothing, dab with eucalyptus oil on a soft cloth.

NOTE

Avoid using eucalyptus oil on delicate fabrics, such as silk. Test by dabbing on a small, inconspicuous area of the fabric before using.

MATTRESS

See individual entries for fabrics and stains.

See also **Wet bed**

MAYONNAISE

Scrape away the excess, then treat as for **Grease**.

MEDICATIONS

Fabric

Flush the area immediately with cold water, then apply a paste of borax and water for a few hours before washing as usual.

- Dab with half a lemon dipped in salt, leave to dry, and then wash as usual.

Carpet

Spray with soda water, and press toweling into the area to absorb as much liquid as possible. If a stain remains, sprinkle with borax or baking soda, leave to dry, and then vacuum up.

MICROWAVE OVEN

To clean out your microwave oven, use *Marvelous Microwave Cleaner* (for recipe see p37).

MILDEW

See **Mold**

MILK PRODUCTS

Fabric

Soak the item in cold water before washing as usual.

Carpet

Mop up with a towel, then sprinkle with baking soda. It will absorb the milk and so prevent any stale milk odor. Once the powder has dried completely, vacuum up.

MIRRORS

To remove paint from mirrors, rub with hot white vinegar on a cloth.

For general mirror cleaning, use a solution of equal parts of white vinegar and water. Spray onto the glass and wipe clean.

MOLD

Fabric

Sometimes mold or mildew will come away simply with a good brush. Use a toothbrush and rub briskly over the affected area. Then wash in warm water.

You can also soak the fabric overnight in a solution of ½ cup of white vinegar and 1 tablespoon of salt. Then wash in warm water, and dry in the fresh air.

✸ Walls and ceilings

Wash with a solution of ¼ cup of baking soda and 1 cup of hot water, then wipe with hot white vinegar. If stains are stubborn, rub with baking soda on a damp cloth.

MUD

Mud stains on flooring will usually come away with a simple wipe with a wet cloth or mop. For stains on clothing rinse under cold running water, then apply a paste of cream of tartar and water. Leave to dry, then wash as usual.

✸ Carpet

Sprinkle liberally with baking soda. If the carpet has a long pile, gently rub the powder in with a soft brush. Leave for an hour, then vacuum up.

MULBERRIES

To remove the deep red stain of fresh mulberries, use unripe, green mulberries. Rub them into the stain, and then wash as usual.

See also **Fruit**

MUSTARD

See **Barbecue sauce**

A–Z

N

NAIL POLISH

Look out for a commercially available organic nail polish remover. To remove nail polish from fabric or carpet, blot up as much as possible, and test the remover on an inconspicuous spot before using it to dab at the stain.

NAPPIES

See **Diapers**

NECK GRIME

See **Collars**

NEWSPRINT

Dab glycerin on the area to loosen the print, then sponge with a solution of ¼ cup of white vinegar and 1 cup of water.

NICOTINE

To remove nicotine from fabric, dab with eucalyptus oil on a clean cloth. Eucalyptus oil rubbed onto nicotine-stained fingers will remove the stain.

NONSTICK PANS

To remove stains, bring to a boil a mixture of 1 tablespoon of baking soda and 1 cup of water. Discard, and rinse pan well.

NYLON

To treat the yellowing stains on nylon, soak it in a bucket of warm water with ¼ cup of baking soda added. Then wash as usual and allow to drip-dry away from direct sunlight.

NYLON SHOWER CURTAIN

See **Shower curtain**

O

OIL

See **Grease**

ORANGE JUICE

See **Fruit**

OVEN

Baking soda will work wonders on your oven, no matter how much in need of a clean it might be. Baking soda sprinkled on a clean, damp cloth and wiped over the inside of the oven will remove most bits. To clean up food spills, sprinkle with plenty of salt while the oven is still warm, and clear away. Then wipe with baking soda on a damp cloth.

To give your oven a good clean, use *Amazing Oven Cleaner* (for recipe see p37).

See also **Microwave oven**

P

PAINT

Hot white vinegar will remove paint spots from windows. Rub with a clean cloth dipped in the vinegar until the spots come away. It is also effective for cleaning up paint brushes: simply soak them in hot white vinegar for a couple of hours before washing in warm soapy water. Then rinse clean.

PAINTED DOORS

Most marks and smudges on painted doors will come away if you wipe them with either white vinegar or baking soda on a damp, clean cloth.

PATENT LEATHER

To remove spotting from patent leather, and to make the surface shine, dab with a little milk on a clean cloth and then buff up.

☛ Petroleum jelly smeared onto patent leather and then wiped away will also clean it and keep it in good condition.

PEACH

See **Fruit**

PEN

See **Ballpoint pen**

PENCIL

You can use an eraser to remove pencil marks from any surface, not just paper.

PERFUME

Flush immediately with soda water, mopping up the liquid with toweling. If the stain persists, dab with glycerin and leave for two hours before either washing as usual or sponging away with warm water.

PERMANENT MARKER

Because these inks are meant to be permanent, they will understandably be difficult to remove. But you'll find that at least some of the stain will come away if you soak the fabric overnight in 1 cup of hot water with 2 tablespoons of borax added. Then wash as usual.

PERSPIRATION

Perspiration marks will come away if you soak clothing in a solution of ¼ cup of baking soda and a bucket of warm water. Soak for a couple of hours before washing as usual.

☛ Warmed white vinegar will also do the trick: soak the affected area for an hour, then wash as usual.

PET FECES

Scrape away any solids, then sponge well with a solution of ¼ cup of white vinegar and 2 cups of water. Sprinkle the area with baking soda, leave for a few hours, and vacuum clean.

✸ PEWTER

To treat any corrosion on pewter pieces, mix together 1 cup of white vinegar, 1 cup of flour, and ½ cup of salt. Apply this thick paste to the affected area and leave for a few hours. Rinse thoroughly, and wipe dry with a clean cloth.

✸ PICKLES

See **Barbecue Sauce**

✸ PLASTIC CURTAINS

See **Shower curtain**

✸ PLASTIC FURNITURE

Stains will come clean if you rub with a paste of baking soda and water. Then wipe down with a clean, damp cloth.

☛ You could also use *All-purpose Cleaner* (for recipe see p49).

✸ PORCELAIN

Rub with baking soda on a damp cloth to remove stains from porcelain sinks and bathtubs.

✸ POTS

To remove burned food stuck to the bottom of

pots, bring a mixture of 1 tablespoon of baking soda and 1 cup of water to a boil in the pot. Discard, and rinse pot well. Repeat if needed.

PRINT

See **Newsprint**

R

RED WINE

Without delay, pour white wine over the stained area. Then mop up with toweling before dousing with soda water. Mop up again using fresh toweling, and repeat the soda water process until all traces of the stain have disappeared. Apply soda water, immediately if you have no white wine.

- For smaller spills, sprinkle salt liberally over the spill. Leave until the liquid is absorbed, then vacuum up.
- If a stain persists, make a paste with a tablespoon each of borax and baking soda together with water, and rub into the stained area. Leave for half an hour, then rinse with clean water.

See also **Alcohol**

✷ REFRIGERATOR

To clean the enamel exterior of your refrigerator, simply wipe down with a cloth dipped in white vinegar.

With a cloth dipped in baking soda, rub away the brownish marks that can form on refrigerators. Then wipe down with white vinegar. For an extra good clean, use *Zingy Appliance Blend* (for recipe see p30).

✷ ROASTING PAN

For those burned bits that won't come away from the bottom of the pan, spread a paste of baking soda and water — be generous — and leave the pan overnight. Discard all that comes away, and wash the pan in hot water.

✷ RUGS

See **Carpet**

✷ RUST

✷ Fabric

Sprinkle the stained area with a generous heap of salt, and then rub in a few tablespoons of lemon juice. Place the garment in the sun, reapplying lemon juice whenever it dries out. Later, rinse the garment.

☛ Sprinkle the stained area with cream of tartar, then dip fabric into hot water.

✷ Carpet

Sponge the rusted area with salt and lemon juice.

S

SALAD DRESSING

Scrape away the excess, then treat as for **Grease.**

SAP

Dab eucalyptus oil onto the soiled area until the sap is removed. Then wash as usual.

NOTE

Avoid using euclayptus oil on delicate fabrics, such as silk. Test by dabbing on a small, inconspicuous area of the fabric before using.

SCRATCHES

Remove scratches on timber furniture by rubbing in a mixture of lemon juice and vegetable oil in equal quantities.

SCUFF MARKS

For scuff marks on flooring, rub with a little eucalyptus oil on a clean cloth.

SEMEN

Fabric

Soak in a bucket of water with 1/4 cup of borax dissolved in 1/2 cup of hot water. Then wash as usual.

Mattress

Sponge with a solution of 1 tablespoon of borax dissolved in a little hot water and added to 2 cups of water. Then sponge again with warm water. Allow to dry thoroughly.

SHEEPSKIN RUG

Rub a little baking soda into the stain and leave before brushing away or shaking it out.

For more stubborn stains, sponge with a solution of 1 tablespoon of borax in 1 cup of warm water. Then sponge with a cloth dipped in white vinegar, and dry the rug outdoors.

SHIRT COLLARS

See **Collars**

SHOE POLISH

Scrape off as much as possible, then dab with eucalyptus oil on a clean cloth. Turn the cloth as polish comes away so as not to re-soil the area.

☛ You could also try glycerin, dabbed on with a clean cloth in the same way, before washing the item as usual.

SHOWER CURTAIN

To remove both mildew stains and soap buildup, scrub nylon shower curtains with warm white vinegar. Then wash in warm, soapy water, rinse, and hang out to dry in the fresh air.

You can help prevent the growth of mildew on shower curtains by soaking them in a tub of water to which you've added 1 cup of salt.

SHOWER TILES

To clean away soap build-up stains, scrub tiles with a small brush dipped in baking soda and a little water. Then wipe over with white vinegar on a clean, damp cloth.

SILVERWARE

Line your kitchen sink with aluminum foil, fill the sink with very hot water, and add either 1 tablespoon of salt and 1 tablespoon of baking soda or 2 tablespoons of cream of tartar. Then immerse the silverware, leaving it to soak for up to half an hour. You'll find that the tarnish will come away during this time. Remove the pieces, rinse them well, and dry with a soft cloth.

SINK

Baking soda on a soft, damp cloth will remove most stains from sinks, whether they are made of stainless steel or porcelain.

SKIN

See **Hands**

SOFT DRINK

✹ Fabric

Run cold water through the fabric as soon as possible after the spill, then soak in a solution of ½ cup of white vinegar in a bucket of water to remove all traces of the soft drink. Wash as usual.

☛ Spray soda water over the affected area, and dab at it with a clean cloth or napkin until all traces of the stain have been removed.

✹ Carpet

Spray with soda water, and press toweling into the area to absorb all the liquid. If a stain remains, dab at it with a cloth dipped in a solution of 1 tablespoon of borax and ½ cup of water.

See also **Fruit**

✹ SOCKS

Socks can become very grubby indeed. Soak very dirty ones for several hours in salty water, as hot as possible for the sock material. Then wash as usual in water, again as hot as possible for the sock material.

SOOT

✹ Fabric

Use adhesive tape (sticky tape) to remove as much of the soot as you can. Avoid rubbing —

this will spread the stain. Dab with eucalyptus oil, leave for an hour, and then wash as usual.

✷ Carpet

Vacuum up as much as possible, and then sprinkle the area with salt to absorb the soot. Vacuum again. If a stain remains, sponge with warm water.

NOTE

Avoid using eucalyptus oil on delicate fabrics, such as silk. Test by dabbing on a small, inconspicuous area of the fabric or carpet before using.

SOUP

The way you treat the stain will depend on the ingredients of the soup. However, soda water is a good general treatment.Spray the area with soda water, and use a towel to mop up the liquid. If a stain remains after this treatment, soak in, or sponge with, 1/4 cup of borax dissolved in 1/2 cup of hot water and added to a bucket of water.

A–Z

SOUR CREAM

See **Milk products**

✷ STAINLESS STEEL

Remove all sorts of stains and marks from stainless steel with baking soda on a damp cloth. Wipe over with white vinegar on a damp cloth to remove any powdery residue.

✸ STOVE TOP

Wipe away spills and spattered food while still fresh. Clean up the stove top with baking soda on a damp cloth, followed by a wipe of white vinegar. A drop of essential oil on any baked-on bits will easily remove them, and leave a lingering fragrance.

☛ Clean the enamel part of the stove top with *Zingy Appliance Blend* (for recipe see p36).

✸ SUEDE

Clean suede by brushing with either a soft brush or a brush with harder bristles, depending on the strength of the material.

✸ SUNSCREEN LOTION

Soak the area in white vinegar until the lotion comes away. Then sponge with water.

✸ SWEAT

See **Perspiration**

✸ SYRUP

✸ Fabric

Soak in a solution of ½ cup of white vinegar and ½ cup of warm water until the syrup stain appears to have dissolved. Then wash as usual.

☛ Add ¼ cup of borax to 2 cups of hot water, and soak in this solution before washing as usual.

✸ Carpet

Sponge with either of the solutions as for fabric, above.

T

TAFFY (TOFFEE)

See **Syrup**

✸ TAPS

See **Chrome**

TEA

✸ Fabric

Rinse with cool water before applying a little lemon or lime juice. Wash as usual.

☛ Soak for a couple of hours in 1 tablespoon of borax dissolved in a little hot water and added to 2 cups of water.

✸ Carpet

Spray with soda water, then mop up the liquid with toweling. Repeat the treatment until all traces of the stain have disappeared.

✸ TEAPOT

The inside of a teapot will come clean if you rub it with damp salt and then rinse with hot water.

☛ Add 1 tablespoon of baking soda to hot water in the teapot. Leave for an hour, then empty out some of the liquid and use a brush to scrub the surface clean. Rinse well.

TEA KETTLE

To clean away the deposit that accumulates inside the kettle, add 1 cup of white vinegar and 1 cup of water and bring to a boil. Empty out and rinse well.

TOMATO

Fabric

Remove tomato stains by soaking in a solution of 1 tablespoon of borax dissolved in a little hot water and added to 2 cups of water.

Carpet

Sponge with the same borax solution used for fabric.

TOYS

For washable toys, wash either by hand or on a gentle cycle using either *General Use Laundry Liquid* (for recipe see p59), or *Gentle Wash for Delicates* (for recipe see p61).

For toys that are unwashable, dust baking soda all over the area to be cleaned, leave for ten minutes, then brush the powder away.

TURMERIC

See **Curry**

U

✸ UNTREATED BRICK

See **Brick**

URINE

Clean up urine on carpet or other flooring by mopping up as much liquid as possible with toweling. Then sprinkle liberally with baking soda, with 3 drops eucalyptus oil or tea tree essential oil added. Allow this to absorb what urine is left. Then vacuum up, and sponge with white vinegar and water to deodorize further and treat the stain.

See also **Wet bed**

V

VEGETABLE OIL

See **Grease**

✸ VINYL

Wash vinyl upholstery with warm water to which you've added some liquid castile soap.

☛ You can also use *Gentle Leather Blend* (for recipe see p50) on vinyl, as well as the *Conditioning Treatment for Leather* (for recipe see p51).

✸ VINYL FLOORING

To clean vinyl flooring, use *Basic Floor Wash* (for recipe see p55), or *Tougher Floor Wash Formula* (for recipe see p56) for a particularly grubby floor.

To remove scuff marks, rub with a little eucalyptus oil on a clean cloth.

✸ VINYL WALLPAPER

Use warmed white vinegar to wipe down vinyl wallpaper.

✸ VOMIT

✸ Fabric

Remove solid matter, then soak in the *Pre-soak Blend* (for recipe see p58) before washing as usual.

✸ Carpet

Remove solid matter. Sprinkle liberally with baking soda, and leave to dry. Vacuum the powder up, then sponge the area with white vinegar.

W

✸ WALLPAPER

Washable wallpaper will come clean if you use liquid castile soap in a bucket of warm water. Follow instructions for cleaning wallpaper on page 46.

For marks on nonwashable wallpaper, rub in a little borax powder and then brush away. A gum eraser will remove marks from wallpaper. You could also use a slice of white bread, rolled into a tight ball, to rub away marks.

WATER MARKS

To remove hard-water marks on pots, add some water to the pot, pop in 1 tablespoon of cream of tartar, and simmer for ten minutes. Rinse and wipe clean.

WAX

Fabric

Step 1: Place the garment in the fridge to harden the wax, and then scrape off as much as you can with a knife.

Step 2: You'll need a roll of brown paper, or a brown paper bag. Put the brown paper on your ironing board, place the stained area of your garment over it, and then cover with more brown paper. With your iron on a low to medium setting, gently press on the stained area. You'll see the brown paper blotting up the wax. Move the paper around, both underneath and over the stained area, and keep pressing with the iron until no more wax is being blotted up.

Step 3: Wash the garment as usual.

✹ Carpet

If the wax is still soft, pack ice cubes around it so that it hardens. Then scrape it away with a knife. To remove any bits, place sheets of absorbent paper on top and use a warm iron to dab at the paper over the stain. Keep turning or replacing the paper as the wax comes away onto it.

✹ WET BED

Soak up as much excess as possible, then mix 1 cup of baking soda with 3 drops of eucalyptus oil or tea tree essential oil and liberally sprinkle over the wet area. Leave to dry, in the sun if possible, and then vacuum up. The baking soda will absorb the fluid and deodorize at the same time. You may have to repeat the process.

WINE

See **Alcohol; Red wine**

✹ WOODEN DOORS

See **Painted doors**

Y

YELLOWING STAINS

To treat yellowing stains on white clothes, soak

for a couple of hours in a solution of ½ cup of lemon juice or white vinegar and a bucket of water. Then wash as usual.

YOGURT

See **Milk products**

Z

ZINC CREAM

Zinc effectively prevents sunburn by protecting the skin, but zinc cream can often get onto clothing, especially when clothing is pulled over the head. For zinc cream smears, sponge with hot white vinegar, and then wash the item as usual.

A guide to weights and measures

USEFUL CONVERSIONS

1/4 teaspoon	1.25 ml
1/2 teaspoon	2.5 ml
1 teaspoon	5 ml
1 Australian tablespoon	20 ml (4 teaspoons)
1 UK/US tablespoon	15 ml (3 teaspoons)

Cup	Imperial	Metric
	1 fl oz	30 ml
1/4	2 fl oz	60 ml
1/3	3 fl oz	90 ml
1/2	4 "	125 ml
2/3	5 "	150 ml
3/4	6 "	180 ml
1	8 "	250 ml
1 1/4	10 "	300 ml
1 1/2	12 "	375 ml
1 2/3	13 "	400 ml
1 3/4	14 "	440 ml
2	16 "	500 ml
3	24 "	750 ml
4	32 "	1L

Index

D

E

F

G

H

I

J

K

L

M

N

O

P

R

S

Notes

First published by Apple Press in 2003
Apple Press
7 Greenland Street
London NW1 0ND
U.K.

www.apple-press.com
Reprinted 2004, 2005 (twice), 2006 (twice), 2007 (twice), 2008

ISBN-13: 978 1 84092 419 0

Commissioned by Deborah Nixon
Text: Angela Martin
Designer: Stephanie Doyle
Editor: Avril Janks
Production Manager: Sally Davis
Project Coordinator: Bettina Hodgson

This book was designed and produced by
Lansdowne Publishing Pty Ltd, Sydney, Australia
www.lansdownepublishing.com.au
Cover designed by Apple Press

Set in Frutiger, Giovanni and Mediterano on Quark XPress
Printed in Singapore by Kyodo Printing (Pte) Ltd